THOUGHT, FUNCTION, AND FORM: THE LANGUAGE OF PHYSICS

By

BRADLEY S. TICE

THESIS

Submitted in fulfillment of the requirements of the American College of Metaphysical Theology for the degree of Doctor of Philosophy in Metaphysics

July 1, 2002

This book is a work of non-fiction. Names and places have been changed tp protect the privacy of all individuals. The events and situations are true.

ISBN: 1-4107-7715-4 (e-book)
ISBN: 1-4107-7714-6 (Paperback)

This book is printed on acid free paper.

1stBooks – rev. 01/24/04

Abstract

The two major discoveries in the physical sciences of the 20th Century are Einstein's General Theory of Relativity and Heisenberg's/Schrodinger's/Bohr's Quantum Theory in the area of physics. These two theories will be examined in light of language as a symbol and language as a metaphor for the description of the processes being conceived. The research will consist of a Lexical-Semantic Model of Analysis that will be used to correlate a concept with a mental dictionary (lexical) and the appropriateness of that word meaning (semantics) to the conceptual meaning of the object or processes being defined.

We are suspended in language in such a way that

we cannot say what is up and what is down.*

Niels Bohr

---

* French, A.P. and Kennedy, P.J. (1985) Niels Bohr: A Centenary Volume. Cambridge: Harvard University, page 302.

Contents

Acknowledgements

It has been a long trail to this point, with many unexpected and unhappy delays, dead ends, and false starts. It is not that I have needed the last twenty odd years to gain the knowledge to write this work, but rather the courage to take steps to do so. A journey too long for such a short trip.

x

Preface

I have found the tie that binds a form with a function, a concept with a physical property, and an image with a word to be the driving force of this dissertation. A balance of semiotics, linguistics, and psychology into the processes and manifestations of the conception and symbolic representation of the science of physics is the net result of this study.

American College of Metaphysical Theology

Minneapolis, Minnesota

hereby confers upon

Bradley S. Tice

The degree

Doctor of Philosophy

together with all the rights, privileges and honors and marks of distinction thereto
in consideration of the satisfactory completion of the course of study prescribed in

Metaphysics

in witness whereof. The signature of its chancellor has been affixed the
First Day of August, Two Thousand and Two.

Dr. Jay Wise, Chancellor

Introduction

Language is a human process, a code, that is rule formed and is manifest, outwardly, with a physical action, usually verbally, that uses symbols to represent both abstract and physical meanings. When applied to print, the symbols become artifacts as much as signifiers of the object being signified. When we discuss or describe a process or concept, we are always utilizing existing terminology to describe that process or concept, even though such a process or concept may not 'fit' such existing descriptions. This then is the major task facing the use of language in the field of science.

Because the field of science is a quest as much as a methodological process, the original observations found from this discovery process results in new conceptual models and modifications of language to describe such new paradigms. Two major discoveries in the physical sciences of the 20th Century are Einstein's General Theory of Relativity and Heisenberg's/Schrodinger's/Bohr's Quantum Theory in the area of physics. These two theories will be examined in light of language as a symbol and language as a metaphor for the description of the processes being conceived. From this a general theory of language and science can be formulated as a unifying bases for the two, that melds each to the other by weight of need and by strength of reason. The research component of this dissertation will be the testing of the Lexical-Semantic Model of Analysis in experiments using four different scientific words.

BRADLEY S. TICE

Review of Literature

In reviewing the quantum theory in general the following secondary works were consulted: Borowitz: 1967; work deals with fundamentals of quantum mechanics, Bohm: 1951; classic text on quantum theory, Bockhoff: 1969; basics on quantum theory, De Broglie: 1953; pilot-wave founder on quantum theory, Cropper: 1970; work on quantum physicists, Cassels: 1970; fundamentals of quantum mechanics, D'Espagnat: 1971; foundations of quantum mechanics, Furth: 1970; modern theories in physics, Fong: 1962; basic quantum mechanics, Eisenbud: 1971; foundation of quantum mechanics, Gottfried: 1966; basic text on quantum mechanics, Guillemin: 1968; history of quantum theory, Kemble: 1937; fundamentals of quantum theory, Jauch: 1968; fundamentals of quantum theory, Landau and Lifshits: 1958; quantum theory textbook, Kursunglu: 1962; basics of quantum theory, Kramers: 1958; textbook on quantum mechanics, Merzbacher: 1961; basics on quantum theory, Mandl: 1957; fundamentals of quantum theory, Lipkin: 1973; fundamentals of quantum theory, Powell and Crasemann: 1961; textbook on quantum theory, Park: 1964; basic quantum theory textbook, Messiah: 1963; quantum theory, Ruark and Urey: 1930; fundamentals of physics, Roman: 1960; particle physics, Slater: 1960; atomic structures, Tomonaga: 1962; quantum mechanics, Van der

Waerden: 1967; quantum science bibliography, Ziock: 1969; fundamentals of quantum theory, Ziman: 1969; advanced quantum theory, Wesley: 1996; classical [early] quantum theory, Wieder: 1973; foundations of quantum theory, Wentzel: 1949; quantum theory and fields, Alonso and Valk: 1973; applications of quantum theory, Afriat and Selleri: 1999; EPR paradox, Hartkamper and Neumann: 1974; ordered linear spaces, Alder: 1995; quaternionics, Bernstein: 1991; quantum scientists, Bethe and Jackiw: 1964; intermediate level quantum textbook, Blinder: 1974; foundations of quantum theory, Bockhoff: 1976; quantum theory, Bohm: 1979; quantum mechanics textbook, Ludwig: 1983, Tannoudji, Diu, and Laloe: 1977, Bud: 1997; measurement and objectivity, Brandt and Dahmen: 1995; pictures of quantum worlds, Duffey: 1992; quantum states, Mann and Revzen: 1996; EPR conference, and Davydov: 1985; quantum mechanics. These are general works on quantum mechanics and quantum theory. They have been used for a general review of the literature.

A more specific area of review was done with the following works: Watanabe: 1983; reductionism in science, Maxwell: 1995; philosopher and quantum theory, Lim and Ald-Shukor: 1998, Black, Pilloff, Sinclair, Nieto, and Scully: 1991; quantum theory, Barut, Feranchuk, Shnir, Tounilchik, and Belarus: 1995; quantum systems, Sucher: 1995; concept of potential, Doebner, Scherer, and Schroeck:

1993; classical and quantum systems, Bub: 1993; measurement and objectivity, Olkhovsky: 1998; time as an observable, Klauder: 1999; metrical properties of quantum theory, Agazzi: 1978; complementarity, Tarozzi and Aluyuvander: 1998, Hilgevoord and Uffink: 1990; new uncertainty principle, Bitsakis: 1985; causality and locality, Tsipis, Popov, Herschbach, and Avery: 1996; new methods in quantum theory, Urigu: 1993; entropy in information, Lofgren: 1993, Griffiths: 1994; linguistic realism, Kochen: 1985; new view of quantum theory, Bell: 1987; hidden variables, Jammer: 1974; philosophy of quantum theory, Hund: 1967; history of quantum theory, Hendry: 1984; Bohr-Pauli theories, Sopka: 1980; America 1920-1935 quantum theories, Heisenberg: 1930;

classic text on quantum theory, and Bunge: 1967; quantum theory and reality. These works are more specific to the question of the philosophies found in quantum theory and are the next step towards the language of science.

Areas specific to language and science are as follows: Lofgren: 1992; linguistic realism, Aggazzi: 1988; complementarity, Beller: 1998; feminist interpretation of physics, Bergstein: 1972; ordinary language and quantum physics, Cahn: 1998; quantum theory in everyday life, Englert: 1996; visibility and information, Faye: 1991; Bohr's life, French: 1985; Bohr's life, Gribbin: 1984 & 1995; Schrodinger's cat and kittens, Haroche: 1998; decoherence, Innis: 1985;

semiotics, Kieffer: 1986; linguistics of german literature, Levy-Leblond: 1988; quantum theory and language, Noth: 1995; semiotics, Popper: 1956; philosophy, Saussure: 1916; linguistics, Schommers: 1995; symbols and signs in quantum theory, Shannon: 1949; information theory, Vvygotsky: 1962; language and thought, and Yam: 1997; quantum theory experiments [Schrodinger]. These works reflect the backbone of research for this research topic and are directly related to the language of science.

As a rule most works deal with language as a subfield of science and not as a primary role in the area of descriptive sciences. Usually language is also treated as a subfield of the philosophy of science and not as a primary field of study of the sciences. This has not had a limiting effect on the types of materials I was able to obtain as they more than matched the needs of this dissertation topic and have been used widely and with great effect to support the research topic. The excellent collections at San Jose State University and Stanford University aided greatly in this literature search.

Quantum Mechanics

Time is a linear track for mortal man and ideas and developments have a 'chronological' order that forms a timeline of events that is called history. The history of quantum theory is defined by a 101 year old history that starts with Max Planck's publication in the year 1900. This timeline is done as a concise map of events that has formed the history of quantum science.

Most historians of science place the beginnings of quantum theory at 1900 with publication of Max Planck's formula for the distribution of energy as a function of frequency in black-body radiation (Davis and Betts, 1994: 1). This concept that light

13

can be emitted or absorbed only in the form of certain discrete energy packages is a development from earlier studies by Boltzmann, Maxwell, Gibbs and others on the statistical description of the thermal properties of material bodies (Gamow, 1966: 6). In 1905 Einstein proposed that light itself was quantized, that light consists of particles localized in space, that was later termed photons (Schommers, 1995: 18).

Next came three models of the atom. Rutherford's (1911) scattering experiments showed a 'planetary' picture of atoms with electrons moving around the nucleus like a planet around the sun (Schommers, 1995: 20-21). Bohr's (1913) model improved on the Ruthford model in that the quantum picture was incorporated by the inclusion of two postulates: the postulate of stationary states and the frequency postulate (Schommers, 1995: 21-22). Sommerfeld's (1916) model included relativistic effects into Bohr's model (Schommers, 1995: 23). Bohr's Correspondence Principle (1920) proved that quantum rules correspond to classical rules of physics. De Broglie's (1923) Wave-Particle Dualism proposed that a particle could also act as a wave (Schommers, 1995: 26-27).

Heisenberg's (1925) Matrix Mechanics makes the operators time dependent (Schommers, 1995: 33). Schrodinger's (1926) Wave Mechanics is the equivalent of Heisenberg's but the time dependence is transferred to the wave functions (Schommers, 1995: 30-33). Born's (1926) Probability Interpretation of the wave function interpreted the individual objects recorded by equipment during an experiment as being true 'corpuscles' (Schommers, 1995: 36-37). Heisenberg's (1927) Uncertainty Relations states that the more marked a wave character becomes, the more vague the corpuscular aspect becomes and visa versa (Schommers, 1995: 44-45). Bohr's Complementary Principle (1927) states that in nature both light and elementary particles can appear as waves or corpuscles depending on the type of experiment performed on them (Schommers, 1995: 44-45).

There are only two theories proposed after 1927 that have some merit towards thinking about quantum theory; Bohm's (1952, 1966) 'Hidden Variables' theory and Everett III (1957) 'Many-Worlds' theory. The major ideas of quantum theory were formed in the years 1925-1926 by Schrodinger and Heisenberg.

BRADLEY S. TICE

A Question of Language

An ideal area to start with questions of language is in the Sturm and Drang period in German philosophy. An insightful passage from Kieffer's book **The Storm and Stress of Language** is on the philosophical implications of language of Hamann (Kieffer, 1986: 13).

> In implying vision, the chief mode of sensual perception, to be dependent on speech, this statement both summarizes Hamann's linguistic thought and provocatively counterpoises it to a traditional view that was shared by all the debaters on the origin of language. The traditional view was that language operates on the undifferentiated mass of our perceptions, organizing it into things belonging to general types, and thereby reducing it to a form manageable by reason. From this perspective, perception is anterior to language in the technical

hierarchy of mental functions. Acting on his conviction about the total linguistically of the world. Hamann now counters that language not only precesses perceptions of knowledge but also provides the occasion for perceiving to commence; language is originating as well as original (Keiffer, 1985: 13).

Kieffer also addresses Schiller's interest in language in the later stages of the Sturm and Drang movement.

It seems that psychology appealed to the student Schiller more than any other subject, because he chose psychophysiological topics for his two dissertations (1779, 1780). To be sure, these academic writings contain nothing noteworthy on language. But it would be surprising if they did, for he was pursuing a medical degree and was constrained from wandering too far afield from empirical data. Soon after abandoning his medical career, however, he gave further expression to his psychological interests in "Theosophie des Julius"; in exploring the relations between mental and physical experience here, he includes fairly close consideration of language as a semiotic system. What he has to say reflects solid knowledge of the view that linguistic signs represent primarily the structures of intellect, not reality in itself (Kieffer, 1985: 21).

The notion that language is 'originating' as well as 'original' is seminal to the philosophy of language in that it empowers it with a creative aspect. Language is not only a part of the perceptual system, but, at times, a primary catalyst or functionary for such perceptions. Thus is a central idea to the 'creative' aspect of science, utilizing all human faculties for a common goal, a totality of sensation channelled, guided, recorded, and communicated, in part, by language (Kieffer, 1985: 12-13).

The question of language as it applies to the concept of quantum theory is really the degree of clarity of descriptions used to enunciate the process being described. The level of ambiguity is lower when a clear model of the object is used in the process of communication. The use of 'clear' and 'clarity' corresponds to the measure of validity to the conceptual model and the level of that model to be used in the communication pathway. In other words, clarity has a quality of robustness for the sign and the signifier of that sign.

Along with the clarity of description comes the accuracy of that description to that process being described. In chemistry the term chemical bond is used to describe the process of a chemical association (Tice, 1997d & 1997e). In a paper presented at the National meeting of the American Chemical Society in Las Vegas, Nevada, I argued that such a term should reflect the action, or verb quality, aspect of the process being described and hence, should be known as a **chemical association** rather than a chemical bond (Tice, 1997c and Appendix).

This may seem a tautology; a needless repetition of an idea, statement, or word, but for the precise use of the terms in context to the physical processes attributed to them. A bond signifies a more 'physical' tie or union than an association.

An association is a relational tie rather than a physical; i.e. cementing or fusing, type of union. To the unaware such terms are interchangeable but when taken in context to specific traits to be expressed, the chemical bond is indeed a chemical association. A similar analogy can be made for the term 'dry' as in to 'let dry' or remove liquid from a substance so that it is 'liquidless'. An example of this is that wet paint 'drys' but cement 'cures', but more often than not we would say that the cement must 'dry', a process that is actually not so much the removal of liquid but the assimilation of molecules to a solid state. This may seem simple but to a materials science expert the difference between 'drying' and 'curing' is night and day. The same can be said for a chemical bond and a chemical association.

This is not a new problem and it effects all fields of study. This can also effect our concepts of how language is to be used outside the human sphere, in regards to non-human objects relating to language (Tice, 1997a).

It is not just terminology that embattles science, but the very nature of what makes up a science can be called into question as can be illustrated by the field of optics. Newton's magnum opus on optics, titled <u>Opticks</u> (1704), was summation of his work of light and that colors were refractions of white light, through a prism, and that this ablation of light was the cornerstone of modern optical physics (Perkowitz, 1996: 57). It was not until the Romantic Movement that a 'backlash' presented itself to Newton's analysis of light, as can be seen by the poet William Blake's rendering of Newton as a seated classically nude figure intently drawing with a compass on parchment, unaware of the world around him [original in the Tate Museum, London, England] (Perkowitz, 1996: 55-56).

This 'imposition of rationality' on the world by Newton was summed up by Blake as being an 'unspiritual act' (Perkowitz, 1996: 46). It would be Johann Wolfgand von Goethe in his Theory of Colors, that was inspired by Newton's work, that would, as explained by Werner Heisenberg, "return to the descriptive science; a science which is interested only in the visible natural phenomena, not in experiments which produce artificial new effects" (Heisenberg, 1983a: 11). What Goethe did was rather than placing a prism in a sunbeam's path, he held it to his eye and saw effects different from Newton's observations, from which he formed theory of color different than Newton's mechanistic observations (Perkowitz, 1996: 55). The English translation of Goethe's Theory of Colours (1840) has the translator, Charles Lock Eastlake, makes some salient comments on Goethe's work:

> There can be no doubt, however, that much of the opposition Goethe met with

was to be attributed to the manner as well as to the substance of his statements. Had he [Goethe] contented himself with merely detailing his experiments and showing their application to the laws of chromatic harmony, leaving it to others to reconcile them as they could with the pre-established system, or even to doubt in conclusions, he would have enjoyed the credit he deserved for the accuracy and the utility of his investigations (Goethe, 1967: ix).

In other words, science can be divided into a process of 'holistic' analysis based only on visible behavior or by reductionist analysis of the parts of the whole.

Of interest also is the universality and evolution of basic color terms. Berlin and Kay (1969) have found that the referents for basic color terms of all languages appear to be drawn from a set of eleven perceptual categories, and that these categories become encoded in the history of a given language in a partially fixed order (Berlin and Kay, 1969: 4-5). These eleven color categories are: white, black, red, green, yellow, blue, brown, purple, pink, orange, and grey (Berlin and Kay, 1969: 2). Of historical note is that the British political statesman and Homeric scholar, William Gladstone produced one of the earliest works on the evolution of color vocabulary in his "Studies on Homer and the Homeric Age" (1858) and finds in this work an uncertainty and inconsistency in the application of color names and notes that the Greeks of Homeric times had no clear notion of color whatsoever (Berlin and Kay, 1969: 134-135).

Brown's <u>Composition of Scientific Words</u> (1956) makes note of how scientific words come into being and why (Brown, 1956). Simplicity is a desirable quality, for it satisfies a primary psychological urge and because it promotes clarity and definiteness we expect others to exemplify it in their creations and communications (Brown, 1956: 34). The nature of the material to be named and its specificity may be a controlling factor in the construction of names (Brown, 1956: 35). Words that sound well appeal to the esthetic appetite of the mind as euphonious words are desirable above those that are not and should be noted by word coiners (Brown, 1956: 35). Single or grouped words are the most effective if they are also mnemonically adhesive and that they use known roots and stems that have been used repeatedly in other combinations (Brown, 1956: 35).

A system for biology uses two Latin or latinized words that constitute a scientific plant or animal species and was developed by Linnaeus (1707-1778) (Brown, 1956: 36). Carolus Linnaeus would use the generative system of plants as a hallmark of classification (Boorstin, 1983:438). Linnaeus developed from John Ray's (1627?-1705) concept of 'species' that was "a set of individuals who gave rise through reproduction to new individuals similar to themselves" (Boorstin,

1983: 434). Scientific terms originate in three ways:

1.) Adoption from existing languages, i.e. Greek and Latin.

2.) Composition by compounding and affixation.

3.) Arbitrary creation without use of evident, antecedent root or stem material.

David Bohm noted that the problem of 'understanding' starts with language. Bohm noticed that both Niels Bohr and Albert Einstein were divided on the theory of quantum mechanics by a common thought that was separated by a common language (Peat, 1997: 248). What was the problem, thought Bohm, was the use of language and the corresponding symbolism, signs and thought common, ie. informal speech, language produced between two speakers. Bohm notes:

> While physicists define their mathematical language with great care, ordinary spoken language is normally taken for granted (Peat, 1997: 248).

From this Bohm came to believe that the trouble with the common forms of communicating physics had there origins within the language itself (Peat, 1997: 248). After this revelation Bohm set about learning about language and languages. From this Bohm ascertained that quantum theory denies independent existing objects and properties, which means that there can be no 'it' or 'the' in the quantum world (Peat, 1997: 248). In Indo-European languages, the use of nouns are paramount and speaker's of such languages think in terms of categories and objects in interaction (Peat, 1997: 250). Bohm believed that perception and communication were indivisible, which means that people also perceive a world composed of localized objects in interaction (Peat, 1997: 250).

The interconnection of language and the world-view had earlier been addressed by Benjamin Lee Whorf (Carroll, 1956) and L.S. Vygotski (Vygotsky, 1962). Bohm, who seemed unfamiliar with the work of either Whorf or Vygotski, continued with his language research and developed a hypothetical verb-based language that he called 'rheomode' (Peat, 1997: 250). Bohm believed that his 'rheomode' language would result in a form of communication that would bind our perceptions with our language and would produce a more clear form of communicating physics (Peat, 1997: 250)[1]. Bohm would ultimately address the duality found in the concept of 'reality'. For scientists the concept of reality is that being 'real' is objective, independent, autonomous, localized in space and time.

---

1 The 'rheomode' language paper is reproduced in David Bohm's Wholeness and the Implicate Order (London: Routledge and Kegan Paul, 1980).

Reality can also be addressed by idealists who claim that reality has origins in the 'universal' mind (Peat, 1997: 259). Bohm's approach transcended this philosophical duality by making a fundamental distinction between 'reality' and what he termed "that which is" [objective] or "all that is" [holistic] (Peat, 1997: 259).

A Philosophy of Science

As P.A.M. Dirac states in his **The Principles of Quantum Mechanics** about how the world operates: "Her [nature] fundamental laws do not govern the world as it appears in our mental picture in any very direct way, but instead they control a substratum of which we cannot form a mental picture without introducing irrelevancies."(Dirac, 1930: Preface). It is interesting to note that Dirac ascribes the concept and terminology of 'direct' in the manner of human perception of the workings of the world. This is because the 'mechanistic' approach developed by Newton and expanded by Einstein's Theory of Relativity is considered to be a linear process, i.e. A is directly observed by B that is observed next to C, etc., and that clear, or at least logical connections, are the result of this type of perception. Dirac also mentions the power of mathematics in his work: "Mathematics is the tool specially suited for dealing with abstract concepts of any kind and there is no limit to its power in this field."(Dirac, 1930: vi). Although mathematics is the most idea language for abstract notions, although artists of various media may argue otherwise, the limitlessness of such a language can be called into question. Mathematics is not a 'rosetta stone' for all human thought and such concepts as 'infinities' and even non-linear operations are less than ideally realized using mathematics than say, philosophy or, in the case of non-linear operations, graphs and notation systems. It must be realized that this work was published in 1930 but much of the thinking expoused by Dirac is still the main stream of mathematical thought today.

Physical laws in science are by there very nature descriptive. As stated from Swartz:

> Although a physical law may in fact apply to no or few items or events in the world, physical laws are not to be thought of as descriptions of specific items or events. Such generality is secured by requiring that the only terms (other than logical and mathematical ones) that may properly figure in a physical law are ones that are purely descriptive (Swartz, 1985: 14-15).

Werner Heisenberg furthers this concept by reducing the effect of quantum theory to the then existing scientific paradigms:

> But this new situation in quantum theory does not necessarily question the traditional method in science; it only questions the assumption that concepts and mathematical constructs can simply be taken from experience (Heisenberg, 1983: 11).

From James Jeans' **Physics and Philosophy**.

> When two hypotheses are possible, we provisionally choose that which our minds adjudge to be the simpler, on the supposition that this is the more likely to lead in the direction of the truth (Jeans, 1958: 183).

This statement must be balanced when taken with the 'weight' of the balance of the concept of 'simpler' as it relates to new theories. Quantum science was not as simple as classical Newtonian science. It was accepted, by most scientists, as an accurate explanation of the physical processes being examined at the time and not a 'rule of thumb' rubric for scientific acceptance of a theory.

David Marr's **Vision:** (Marr, 1982).

In other words, vision is the process of discovering from images what is present in the world, and where it is...The study of vision must therefore include not only the study of how to extract from images the various aspects of the world that are useful to us, but also an inquiry into the nature of the internal representations by which we capture this information and thus make it available as a basis for decisions about our thoughts and actions (Marr, 1982: 3).

Marr is connecting the act of vision with the cognitive process from that act of 'seeing'. This is fundamental to the understanding of the 'concepts' behind the visual environment that is the foundation for visual stimuli and the resulting conceptual models that form in our brains of that world. This is the key to the shapes and forms that surrounds our mental images and reflects the standards that unify the world to all of those who question the world about them. Marr continues:

> Most of the phenomena that are central to us as human beings – the mysteries of life and evolution, of perception and feeling and thought – are primarily phenomena of information processing, and if we are ever to understand them fully, our thinking about them must include this perspective (Marr, 1982: 4).

The act of seeing and the act of processing what was seen is a matter of cognitive actions that interact with the physical process of sight. From this an information model of input/output can be defined as follows: the input of an image through vision or sight will produce a corresponding output of language descriptions on what was seen. From this model a bases of empirical models can be defined as a bases for scientific inquiry and the language of that inquiry.

The visual information that is the bases of empirical science is the primary or core area of perceptual input of the human faculties for such scientific reasoning. The visual images are those that are stored and used by the brain for the design by the mind to achieve a conceptual model of abstract and tangible thoughts. Such 'visual thinking' has been attributed to Einstein's mind as related in the following passage by West.

From Thomas G. West's **In the Mind's Eye.**

> The late use of language in children, the difficulty in learning foreign languages...may indicate a polarization or displacement in some of the skill from the verbal to another area. That other, enhanced area is without a doubt, in Einstein's case, an extraordinary kind of visual imagery that penetrates his very thought processes. These observations by Gerald Holton, a physicist and a historian of science, seem surprisingly close to those of the neurologists Geschwind and Galaburda. This basis and extent of the presumed associations of visual thinking, creativity, and certain linguistic or mathematical disabilities are still far from clear (West, 1991: 40).

Notice that West uses Einstein as a case for 'visual imagery' in defining a process of the mind. But one should not assume that the mind, in this case Einstein's, is lacking a language component to match the visual centers of the imaginative mind, because Einstein was a great thinker and philosopher and was well equiped to deal with the verbal aspects of scientific thinking as well as the visual component of the human imagination as reflected by the human mind.

The philosopher Karl Popper discusses the importance of language as being a world in of it self and that this world, World III, is the product of human thought and reason. While too abstract to tie into a case for quantum science existing only in the this World III, the concepts and philosophy behind quantum science is that of human language and human concepts. While Popper has opened the intellectual door for inquiry his work is not developed to the point where it can be used to develop or make a case for thought processes of quantum science beyond the simple nature of a general relational aspects of human mind/ human language interaction. The following is from Karl R. Popper's **The Open Universe** (Popper, 1956).

It is interesting to note that Hobbe's belief that the physical world was deterministic preceded Newton's Theory. Newton's magnificent success thus could readily be interpreted as a most impressive corroboration of the determinist doctrine. It seemed that Newton had turned the old determinist programme into reality (Popper, 1956: 26).

Or in other words, 'scientific' determinism could, if at all, follow only from a system of physics which was complete, or comprehensive, in the sense that it would allow the prediction of all kinds of physical events (Popper, 1956: 38).

The method of science depends upon our attempts to describe the world with simple theories: Theories that are complex may become untestable, even if they happen to be true. Science maybe described as the art of systematic over-simplification-the art of discerning what we may with advantage omit (Popper, 1956: 44).

Notice that Popper is using a form of 'reductionism' to give an account of simple scientific theories that form the backbone of scientific hypothesis that may, by way of the ambiguity present in language, allow a form omission and commissions in communicating thoughts about an idea that is presented in 'too simple' a manner. Reduction as a process in physics is addressed in Watanabe (1983) and is found to be a non-tenable process in the discipline of the physical sciences (Watanabe, 1983: 261-264). Popper continues:

> Special relativity, in spite of its prima facie determinist character, cannot therefore be used to support 'scientific' determinism, for two reasons. (1) The predictions demanded by 'scientific' determinism must be interpreted, from the point of view of special relativity itself, as retroctions. (2) Being retroctions, they appear, from the point of view of special relativity, to be computed in the future of the predicted system. Thus they cannot be said to be computed within that system: they do not satisfy the principle of predictability from within (Popper, 1956: 61).

It is not clear in ablating a future process, a retroction, with a predictive value found in relativity as a process 'outside' the system of relativity. As a future function that is both deterministic and a functional part of the continuum of the relativity system they are each a part of the other and hence a 'deterministic science' is the result. Predictions of the future are a part of the past potentials of predictions and so are bound by the process of potentials rather than by space and time. A prediction is a part of itself regardless of whether it's potential is known or not.

Thus, World I, or the physical world maybe taken as the standard example of reality or of existence (Popper, 1956: 116).

> The proposition the truth of which I wish to defend and which seems to me to go a little beyond common sense is that not only are the physical world I and the psychological world 2 real but so also is the abstract world (Popper, 1956: 116).

Popper's view is a strong force for logical examinations of the way in which scientists view the world because the concept of 'real' is not confused with a tangible process, but rather a potential that may or may not be a 'physical' character of the physical world, but is nonetheless a symbolic 'reality' from which the physical world has roots.

> The Cartesian physical universe was a moving clockwork of vortices in which each 'body' or 'part of matter' pushed its neighboring part along, and was pushed along by its neighbor on the other side. Matter alone was to be found in the

physical world, and all space was filled by it. In fact, space too was reduced to matter, since there was no empty space but only the essential spatial extension of matter. And there was only one purely physical mode of causation: all causation was push, or action by contact (Popper, 1956: 135).

My thesis is that, with the higher functions of the human language, a new world emerges: the world of the products of the human mind. I have called it 'World' 3 (Popper, 1956: 154).

Acceptance of Popper's world's is similar to acceptance of the two schools of though about the quantum world. From an article in **The Economist**:

It seems to come down to a choice between accepting that human consciousness is somehow involved in determining the properties of sub-atomic particles; or believing that an infinite number of equally real universes are out there, hidden from human perception (The Economist, 1999: 91).

Upon reflection neither of these hypothesis seems to add much help in answering the questions of the quantum world. Human consciousness is always a factor in observations and deductions and only plays a role in observing, and not the destiny of what is observed. We are humans not gods. Infinite universes are an ugly concept, where are these 'other' universes stored? In someone's old shoe box? Multiple possibilities do not produce multiple universe's. Only unimaginative theories do.

From Max Planck's **Where is Science Going** (Planck, 1932).

Thus the supreme task of the physicist is the discovery of the most general elementary laws from which the world-picture can be deduced logically (Planck, 1932: 10).

When Einstein promulgated his relativity theory much of the enthusiasm with which it was proclaimed was associated with the impression that it constituted a complete overthrough of Newtonian doctrines; where as, as a matter of fact, relativity is an expansion and refinement of Newtonian physics (Planck, 1932: 31).

The use of 'overthrough' is a poor word choice as Planck states that relativity is an expansion of Newtonian physics, not a deconstruction of it. While relativity theory is a new paradigm, it does so not at the complete expense of the existing Newtonian one.

Beyond the achievement of welding space and time together with the mechanical laws of motion, the relativity theory accomplished another and no less important amalgamation. This was the identification of mass with energy. The unification of these two concepts establishes for all equations in physical science the same kind of symmetry as the four coordinates of the space-time continuum, the momentum vector corresponding to the place vector and the energy scalar corresponding to the time scalar (Planck, 1932: 55).

Logic in the purest form, which is mathematics, only coordinates and articulates one truth with another. It gives harmony to the superstructure of science; but it cannot provide the foundation or the building stones (Planck, 1932: 65).

Planck notes that the language of mathematics is descriptive, not constructive of ideas in science. A code rather than what is in the code.

Copernicus discovered nothing. He only formulated, in the shape of a fanciful mental construction, a mass of facts that were already known. He did not add anything to the store of scientific knowledge already in existence (Planck, 1932: 72).

The ideal aim before the mind of the physicist is to understand the external world of reality. But the means which he uses to attain this end are what are known in physical science as measurements, and these give no direct information about external reality. They are only a register or representation of reactions to physical phenomena. As such they contain no explicit information and have to be interpreted (Planck, 1932: 84).

Planck is clear to note that Einstein's Theory of Relativity is an extension of Newtonian Physics, not a break from it, and that such a theory is built on Newtonian fundamentals that where thought up by 1666!

Fundamental ideas of thought are sometimes simple to convey, yet are the crux of the matter, as can be seen from a passage from Gamow: Gamow states:

> The great idea, which was included by Einstein in the foundation of his general theory of curved space, consists of the assumption that the physical space becomes curved in the neighborhood of large masses; the bigger the mass the larger the curvature (Gamow, 1947: 106)

Heisenberg started quantum mechanics with "a brilliant idea: one should try to construct a theory in terms of quantities which are provided by experiment, rather than building it up, as people had done previously, from atomic model which involved many quantities which could not be observed. This amounted to a new philosophy

Dirac said." (Gleick, 1987:72). Heisenberg's 'idea' is not so much a new philosophy, i.e. the very nature of empirical science is that observations are made of experimental processes and that the conclusions reached 'after' the termination or completion of an experiment is when a proposed theory or hypotheses proves true or false, is when the idea is molded into a conceptual framework that gives support to the concept being tested, in as much as he proposed a 'different' aspect to the philosophy of physical sciences that was being proposed at the time, i.e. Niels Bohr's classical atomic model of the atomic world. Dirac is being too broad in giving a variation on a philosophy as being a 'new' philosophy. Words, again, can be misleading to conceptual models of communicating ideas from one person to another, and in this case the trouble is the static use of philosophy rather than the fluid use of hierarchy's of ideas within the domain of the philosophy of science. Bohr's was driven to invent quantum mechanics because the stability of matter required it and that these gross properties of the materials, their color, conductivity and strength, reflect the nature of their quantum mechanical states (Cahn, 1998: 57).

Arthur Pap in his **An Introduction the the Philosophy of Science** states:

This is true especially of statements describing what we believe to be laws of nature: the law of gravitation, the laws of thermodynamics, the laws of chemistry, the laws of heredity-these are all universal prepositions, stating that under specific conditions such and such happens always an everywhere, which have been empirically confirmed (Pap, 1962: 16-17).

But a physical difference that is so small that it is not directly detectable even by means of the most delicate measuring instrument may none the less be indirectly detectable through mathematical deduction of consequences (Pap, 1962: 45).

Pap is incorrect in this analogy as mathematics is the language of observations and hence is constrained by both the nature of the language, mathematics, and the nature of observation. When a mathematical model defines more than a observation, then the hypothesis for that model, or guess, is inaccurate. The language of that model, mathematics, is only a code for that model and not a part of the hypothesis of the model except as a communication system. Mathematics only reflects the nature of the model, not the model itself.

> A substitution instance of a law of logic, such as "if all Negroes have black skin and some Americans are Negroes, then some Americans have black skin", is called a logically true statement (Pap, 1962: 96).
>
> But further, it is just wrong to claim that the Euclidean character of physical shape follows analytically from the very meaning of "measurement of length" insofar as the geometrical concepts are operationally defined, i.e., with reference to the operations of measurement. The most that could be substantiated by this line of reasoning is that the "axiom of free mobility" is such an analytical consequence. A space is said to satisfy this axiom if bodies can be moved in it so as to remain congruent to themselves (Pap, 1962: 121).

Where ever the verification of a physical theory, whether Newtonian or relativistic, involves measurements of length (distances), the assumption of free mobility is innocuous because Euclidean geometry, and a fortiori the axiom of free mobility, is still a correct approximating description of small spaces (Pap, 1962: 122).

The important consideration is that 'measuring' makes sense only with respect to variable properties (Pap, 1962: 125).

This is an overstatement as the need to find a measurable amount in a group of objects may result in all in that group to be of an equal length. This does not reduce the need for such a measurement. A potential for measuring is not solely based on variable properties, but on the need to measure.

As an abbreviation for "measurable determinable property" we shall use the word "magnitude". Length of 3 feet is a quantity. And length is a continuous magnitude, whereas cardinality (the determinable of which particular cardinal numbers are determinate forms) is a discrete magnitude (Pap, 1962: 126).

Whereas there are purely deductive sciences, there are, therefore, no purely inductive sciences. Nevertheless, it is characteristic of an empirical science that it also involves inductive reasoning (Pap, 1962: 139).

The only reason, it will be recalled, why von Mises and Reichenbach felt it necessary to define probabilities as limits was that they postulated that the reference classes (or "collectives") be infinite, and noted that it is meaningless to speak of ratios with an infinite denominator (Pap, 1962: 189).

It might be equally argued that unknown quantities should be assigned as 'indefinite' quantities as they are not known to be infinite, just indefinite in length.

It is, of course, involved in the law of inertial any isolated body is at rest or in uniform motion relative to any inertial system. It is also involved in the postulate of the special theory of relativity that the velocity of light (in vacuo) is the same in all inertial systems. What is meant by an inertial system? Three definitions may be considered:

1. A system relative to which an isolated body either at rest or in uniform motion (Pap, 1962: 294).

2. A system in which no inertial forces such as centrifugal forces, manifest themselves (Pap, 1962: 295).

3. A system that is not accelerated relative to the fixed stars (Pap, 1962: 295).

Einstein concluded that the transformation equations must be modified in such a way that the postulate c=c', which to his mind was dictated by the very meaning of "law of nature", would be satisfied. There resulted the Lorentz-Einstein transformation equations with their well-known consequences that measured lengths and times depend on the state of motion of the coordinate system relative to the measured objects or processes. The new transformation equations are still restricted to inertial coordinate systems, but they unify the laws of mechanics, optics, and electromagnetics as being invariant with respect to the same group of transformations (Pap, 1962: 299-300).

The controversial question is whether this reason determinism in general has broken down in the revolutionary "quantum" theories of the Twentieth Century. Those who draw this conclusion usually refer to Heisenberg's uncertainty principle according to which the product of the "uncertainty" is the position of a subatomic particle and the uncertainty of its simultaneous momentum cannot be less than h, the "quantum of action":

Delta (x) multiplied by Delta (px) greater than equal to h (Pap, 1962: 320).

The question of 'reason determinism' as addressed by Pap (1962) is not so much 'our' knowing positions of subatomic particles but that even the unknown can still be considered a form of 'determinism', i.e. our knowledge of position is irrelevant to whether the fact that the positions of subatomic particles do exist. Our knowledge of 'preordaining' a position is absurd. Human reason is a logical abstract of measured places and times and is still a form of reason even when such measurements cannot be obtained. As determinism is still a form of human conceptual perception to the formulation of space and time to an unobserved event such that is found in the quantum world. They are not diminished by the lack of information about them. They are only less satisfying and not as complete.

Erwin Schrodinger published his now famous 'cat experiment' in <u>Naturwissenschaften</u> in 1935

(Volume 23, page 812) that is, in essence, the following:

A box containing a radioactive source, a detector that records the presence of radioactive particles, a glass bottle containing a poison and a live cat.

As described from Gibbin's **In Search of Schrodinger's Cat** (1984):

> The apparatus in the box is arranged so that the detector is switched on for just long enough so that there is a fifty-fifty chance that one of the atoms in the radioactive material will decay and that the detector will record a particle. If the detector does record such an event, then the glass container is crushed and the cat dies, if not, the the cat lives. We have no way of knowing the out come of this experiment until we open the box to look inside, radioactive decay occurs entirely by

chance and is unpredictable except in a statistical sense (Gibbin, 1984: 203).

An interesting idea presents itself from this thought experiment as related from another book by Gibbin **Schrodinger's Kittens and the Search for Reality** (1995):

And is life a requirement of a conscious observer, in this sense of the term? Would a sophisticated computer be able to collapse the wavefunction by looking into the room? (Gibbin, 1995: 22).

The question of a computer being equated with what we consider 'life', i.e. highly developed organic life forms, is similar to the question raised in my paper (1997a) on Turing's question 'Can a Machine Think?'. As a thought experiment, the use of a computer to 'look' into the box, verses a human, will probably be statistically the same amount of 'interference' as a human, because they both are 'interacting' with the experiment and both suffer from the 'observer's paradox'.

Quantum science as explained by the now famous **Schrodinger's Cat Experiment** is typical of real world allegories that do not fit well as a language of the quantum world as explained by Haroche:

> Quantum experts will object that a cat is a complex and open system which cannot, even at the initial time of this cruel experiment, be described by a wavefunction. The metaphor, nevertheless raises an important question: Why and how does quantum weirdness disappear in large systems (Haroche, 1998: 36).

An example of the misuse of 'everyday' concepts can be found in the following article by Yam (1997). In Yam's article (1997) that discusses Pritchard's, et al, Schrodinger's cat experiment he states:

> By cooling particles with laser beams or by moving them through special cavities, physicists have in the past year created small-scale Schrodinger cats. These "cats" were individual electrons and atoms made to reside in two places simultaneously, and electromagnetic fields excited to vibrate in two different ways at once (Yam, 1997:124).

The notion of a 'small-scale Schrodinger cat' is an oxymoron. A cat is an animal. Made up of a complex system of molecules that form a specific phenotype/ genotype behavior. The 'small-scale Schrodinger cat' that is being presented by the research of Pritchard, et al, is actually not even a 'molecular' level object, but an atom and electrons, the atomic level. Because a cat is an open system phenomena, the allegory of an atom behaving like a cat in Schrodinger's gedanken [thought] experiment is incorrect. A cat is a complex molecular system. The atom is the basic unit of the physical world. The realities of one are not the realities of the other and this notion of a scaled down feline is erroneous to Schrodinger's thought experiment.

Yam is clear to note why such large scale objects do not behave like a single atom:

It also explains why size per se is not the cause of decoherence: large systems, like real-life cats, would never enter a superposition, because all the particles that make up a feline influence a vast number of environmental parameters that make coherence impossible (Yam, 1997: 127).

Quantum super positions must somehow yield outcomes that conform to everyday sense of reality. That leads to circuitous logic: the results seen in the macroscopic world arise out of the quantum world because those results are the ones we see (Yam, 1997: 128).

Some of the problems with quantum mechanics is that it is loaded with the perceived 'burden' of thought of classical physics. There is actually no difficulty in thinking about superposition of a single phenomena. Only previous thought has 'trained' our thinking in such matters of perception, i.e classical physics. As an example from the everyday world we are use to 'seeing' reflections of ourselves using mirrors. Because of

71

our binocular vision, we see only that which is before us. We do not, unaided, see 'behind an object'. If, for a thought experiment, we where observing an individual facing us and behind that individual was a full-sized mirror, i.e. reflecting head to toe the person standing before us. We would 'see' upon looking at the individual in our line of sight, their front, but with the aid of the mirror, also their back.

We would have, upon a single glance, the front and back of a person in our view. We are use to this phenomena, but if we had never used a reflecting device before, i.e. a mirror, we would only be accustomed to a single angle, either front or back, but not both. The mirror in effect becomes an experimental device, and the resulting dual images are received as a front and back 'real time' image of an object. This is, upon reflection, a strange phenomenon, but because we have all shared this phenomena, it becomes common place. Perhaps quantum thinking must also evolve to this point, in order for it to grow past the 'legacy' of classical thinking.

Bergstein's **Quantum Physics and Ordinary Language** (1972) states:

> The reason why so much has been spent trying to retain classical objectivity in quantum physics is obviously the basic psychological condition that the conceptual scheme of ordinary language used when comprehending external phenomena is not easily dispensed with in natural science (Bergstein, 1972: 33).

Bergstein continues:

> It is seen that the describability in ordinary language of the observed phenomena is a basic constituent of physical objectivity. Quantum physics as well as classical physics is necessarily founded on the common ability of observing and describing unambiguously the external phenomena of everyday life (Bergstein, 1972: 36).

Faye (1991) discusses Neils Bohr's similar line of reasoning with 'ordinary' language:

> ...the forms of perception are the preconditions of the possibility of sensory experience as well of the meaning of the ordinary language which we use to communicate this experience, including the theoretical refinements of the ordinary words employed in physics (Faye, 1991: 133).

It is interesting to note that Einstein had a more 'moderate' view of measurements in quantum science as he felt that wavefunctions did not possess a complete 'picture' of the observer-independent physical reality, whereas Bohr and Heisenberg seemed to deny any such reality of atomic phenomena (Goldstein, 1998: 43). In the long run the Bohr-Einstein debate, as it was called, has lead to Einstein's more modest estimation being varified by three basic approaches: decoherent histories, spontaneous localization, and pilot-wave theories (Goldstein, 1998: 43). While these three theories are interesting they do not address the question of language in physics as much as it does the physics itself.

Mara Beller makes an interesting note that the physicists Bohr, Born, Heisenberg and Pauli's philosophical pronouncements on quantum science has lead to the excesses of the postmodernist critique of science (Beller, 1998: 29-34). Postmodernism seems to be the general need for those uniformed in a discipline to make pronouncements about that subject even though such comments are not backed by a 'formal' education. This 'general agreement' by the great unwashed has, of course, had a severe impact on the sciences. Just because a few prominent physicists try philosophical musings in their areas of interest and expertise, at least at the theoretical and experimental level, this does not make them experts at the philosophical or epistemological levels. In Spielberg and Anderson (1987) the following statement is made concerning quantum theory to other fields of study:

> On another level, the idea that there are fundamental limits on the measurement of certain quantities in physics has suggested to workers in other fields that there may be analogous fundamental limits in their own disciplines on quantities to be measured or defined (Spielberg and Anderson, 1987, 223).

This is, of course, the major reason why postmodernism has spread to other fields because of the general feeling that this 'inherent' lack of knowledge to all things or a limit to this knowledge will place barriers to knowledge in the future and has resulted in the disregard for all known knowledge, as if it had no value at all. As noted in Bauer in regards to the limits of science: "That sciences does not have all of the answers does not mean that it has no answers" (Bauer, 1992: 144).

Bohr was a tortured speaker and Heisenberg was an 'all or nothing' philosophical thinker that, along with Bohr, made the question of quantum reality of observer-independent measurements an 'invalid' option. Beller is on target with this article but just because physicists are inconsistent with their philosophical prognostications, it is not clear from that standpoint that others, i.e. everyone else, must feel the need to question the validity of quantum theory or, even worst, ascribe it to other disciplines as well. Now it is well documented that many prominent scientists have said questionable things about the philosophical ramifications of quantum science, i.e. to social and political questions, but this does not excuse the rest of the human race from adding to this situation by adding their own 'questionable' ideas to a very difficult concept. Postmodernism is just a stupid form of modernism where the questions are open to everyone and the resulting mess is the net consequence of stupid people, not stupid science.

BRADLEY S. TICE

The Language of Potentials

A potential must be viewed as both a static, non-active kinematic property, and an active, active kinematic property, to be valid. Does this mean that that a potential can be both? It is usually either one property or the other, but in an 'indeterminate' state in can be factored as both. Why an active potential? Think of butterfly hunting with a net. The butterfly can be taken both 'on the wing', in flight, and stationary, when landed. When thinking of an active environment at the particle level, the late David Bohm's work with 'hidden variables' is the most well known. Although Bohm has written an excellent account of quantum theory (1951) he is best known for his theory of the holomovement, i.e whole or holistic view of the physical universe, that deals with active potentials at the quantum level.

According to Bohm, the major point of the holomovement is that it is represented by movement. Think of a film rather than a photograph. Current quantum theory views the particle world as a static event. Bohm suggests that the particle world be viewed as an action related environment (Sharpe, 1993: 50). Grammar also reflects the objective metaphysics that cultural conditions force upon the western mind. The noun, the indicator of an object, has the primary grammatical role while verbs, which call attention to action, have a secondary status

(Sharpe, 1993: 51). Bohm would like to see the verbs as the primary grammar feature of the inquiry into a quantum environment. The language of movement is the language of the verb. Bohm's basic theory is that the particle world is active, not static, and that our minds must grasp this action oriented world if we are to develop a progressive view of the quantum world. The holomovement is an unbroken, undivided wholeness that is developed from fragmentary 'observations' from that whole world environment.

Bohm's theory deals with what are called 'hidden variables'. These hidden variables are just 'potential' processes that have not been found in the particle or subparticle world of quantum theory. Based on Schrodinger's wave equation that represent continuous trajectories for particles at the subquantum mechanical level, and that hidden variables play a role as a determining factor (Schommers, 1995: 50). Hidden variables are just potentials within the quantum world that exist on a theoretical plain, but with little to go on outside of Bohm's musings. Bohm's theories are more of a philosophy for broader thought of the quantum world rather than a particular physics experiment to 'prove' such hidden variables exist. Probably the best case against hidden variables is von Neumann's theorem showing that quantum theory does not allow hidden variables (Sharpe, 1993: 19).

This is not to discount Bohm's theory of thought. Bohm is keeping an accurate picture of the quantum world in his thinking. Whether such a world can ever be examined to give a 'whole' pattern to the understanding of the quantum world is another matter. Nonetheless Bohm's thinking in the matter should be the very bases of physical sciences as nature is never at rest and neither should be the minds of those that study her [nature].

While examining Bohm's view on language would be better suited in the next chapter, the question of language to Bohm's world will be discussed at present. Bohm is quick to note the very nature of language, grammar; the rules of a language. The noun verses verb statement is precise and correct in defining the needs of language to the discipline of science. Nouns are static and dull, hence proper nouns have the mundane feature of representing names of people, places or things. Verbs are pure jazz: movement and action. The atomic and subatomic world are orchestrations in movement. The melody of particles is played to the rhythm of action. No nouns need apply to describe this world. Bohm's ideas are the centerpiece of thought about the quantum world.

Another view point is the 'consistent-histories' theory that treats quantum mechanics as fundamentally a stochastic or probabilisitic theory, as far as time is concerned, and does not introduce deterministic features to this randomness by the use of 'hidden variables' (Griffiths and Omnes, 1999: 27). The basic task of quantum theory is to use the time-dependent Schrodinger equation, not to generate deterministic orbits, but instead to assign probabilities to quantum histories, sequences of quantum events at a succession of times, in much the same way that classical stochastic theories assign probabilities to sequences of coin tosses or to Brownian motion (Griffiths and Omnes, 1999: 27). In essence, the constant-histories approach takes the concept of measurement as not the basis for interpreting quantum theory, but as measurements that can be analyzed, with other quantum phenomena, in terms of physical processes (Griffiths and Omnes, 1999: 26). Although Griffith and Omnes state that general rules are in place in 'consistent-histories' theories to reduce contradictions and paradoxes in using retrodictions found in consistent-histories applications, such a claim must be balanced with the fact that there is no demonstrable proof that quantum systems actually exhibit irreversible behavior in a thermodynamic sense, and that thermodynamic irreversibility is an effect closely related to quantum decoherence (Griffith and Omnes, 1999: 29 & 31).

Of interest is the term 'consistent-histories' and the 'truth' of terminology as it applies to the concept being addressed, namely do 'stochastic' measures of quantum events reflect a singular nature of a single 'quantum event' in succession or is it a 'time' sensitive mapping of quantum events that has been 'grouped' together in a linear time frame? Also the 'history' aspect of the 'consistent-histories' approach is misleading as most histories deal with singular natures that are traceable over time, i.e. the same object being tracked over a length of time, and this 'begs the question' are these histories being forced into singular events of actually random, non-related quantum events that still can only be addressed as point particles or wave functions of the quantum world? Also the words 'consistent' and 'history' are redundancies in that history deals with chronological timelines of events that are bound by time and space, but also by events before and after each abstraction of the segmented event's histories.

History is a continuum by nature and consistency is just a continuum that is not derivational in nature, i.e. does not change, and that one supports the other in being recordable, consistent, time and space dependent, and connected. There is little in 'consistent-histories' theory that supports these definitions and would be better if it were termed 'perceptual events mapping' theory or 'multiple events' theory or some other moniker that would allow some 'unknown' aspect to be addressed in the theory without destroying the interesting nature of the experimental aspect of

this otherwise well thought out theory.

BRADLEY S. TICE

The Word as Fact

Niels Bohr was both a philosopher and a physicist and his interest in language as it relates to everyday matters and as descriptions of ideas is well known.

> To Bohr, philosophical problems were neither about existence or reality, nor about the structure and limitations of human reason. They were communication problems. They dealt with the general conditions for conceptual communication (French and Kennedy, 1985: 301).

Bohr deals with the communication pathway and the inherent dangers faced with the ambiguity of the signal source and the signal receiver to the signal being sent. Bohr shows great insight in focusing on the language route, rather than the broader areas of language philosophy that are more esoteric, although just as interesting, than the communication model he is describing. In many ways Bohr is pre-dating the work of Claude Shannon in the area of information theory as it applies to language feedback systems (Shannon and Weaver, 1949). Although Bohr is a philosophical analysis and Shannon's is a mathematical one, they both are general models of communication systems that parallel the human activity accurately.

> "What is it we human beings ultimately depend on? We depend on our words. We are suspended in language" (French and Kennedy, 1985: 301).

The idea of being 'suspended' in language is a graceful metaphor for the attributes language affords man in communicating to his fellow man. It is not just a process, but a life sustaining process that is a mark of man and an evolutionary development that separates man from all the other higher life forms on this planet.

The general conditions for the use of language include a law requiring a proper balance between content and form in conceptual communication. When we describe and order experience, we must use a system of concepts. No experience can be understood or communicated without being fixed in a logical frame. The frame-that is, the way we characterize and combine experience – determines what we can talk about and what relationships we can express. We must always be prepared to find that a conceptual framework is too narrow to contain the content we want to press into it. In such a situation we are confronted with a logical disharmony, because we try to speak about something for which our conceptual system has no room. And in efforts to restore harmony, even the frames that are apparently the most solid, those defining our elementary concepts, may prove to be blinders that conceal more fundamental relationships. Yet logical possibilities for extending or generalizing any frame lie like seeds in the presuppositions for using our concepts. The extension enables us to talk about new things and to express new kinds of regularities. (French and Kennedy, 1985: 301).

Deductive reasoning has taught us the significance of the conceptual framework (French and Kennedy, 1985: 301).

The method for analysis of the conceptual framework, i.e. the experiences of life that shape the meaning behind our words, is deductive reasoning, to derive at an answer by subtraction until an answer becomes apparent.

> In addition, mathematics has shown us the
> wealth of unsuspected possibilities for
> conceptual extension or generalization that
> are latent in the way we use our simplest
> words (French and Kennedy, 1985: 301).

Mathematics is the method for this deduction as it

has the power to create or extend the constraints

inherent in the body of words. Mathematics is the

key to clarity, a pure language, that has at it's

heart, the logic of reason. Perhaps the real

question of language is reason and that is why

the 'philosophers' language was always the realm

of logic.

Aage Petersen, in French's and Kennedy's **Niels**

Bohr: A Centenary Volume, sums up Bohr's

philosophy in the following passage:

As far as I can see, the doctrine that we, philosophically speaking, suspended in language, that we depend on our conceptual framework for unambiguous communication, and that the scope of the framework be extended by generalization in the way illustrates in mathematics, forms the general basis of Bohr's philosophy (French and Kennedy, 1985: 302).

Petersen supports this philosophy of Bohr's use of mathematics as a 'conceptual razor' to cut to the core of the concept and is thus the language of science. This is not new as most scientists have strived for a pure, or rational, language throughout history to aid in the communication of such information. The historian, and Librarian of Congress, Daniel J. Boorstin states in his **The Discoverers**:

The Royal Society, hoping to accomplish this, therefore, "exacted from all their members, a close naked, natural way of speaking; positive expressions; clear senses; a native easiness: bringing all things as near the mathematical plainness, as they can: preferring the language of Artizans, Countrymen, and Merchants, before that, of Wits, or Scholars" (Boorstin, 1983: 395).

It is clear from the onset of the Royal Society that a specific form of language, in this case the written form, was to embody the 'plain style or mathematics' in substance, if not in form. This form of language is with science publications to this day.

The chief characteristic of the sort of description we seek both in science and in practical life is objectivity. In Bohr's usage, an objective message was an unambiguous message, one that could not be misunderstood. If our communications are to be understood, their content must be clearly delineated. There must be, so to speak, a partition between the subject which communicates and the object which is the content of the communication. This partition is indispensable in every objective description, and Bohr saw in it the core of the problem of knowledge (French and Kennedy, 1985: 302).

Bohr's need for an 'unambiguous' signal for the communication pathway makes it the primary area of coding for an objective account of the observable world. This parallels signal noise, or rather the amount of signal noise, in the mathematical models of Shannon Theory and is an accurate predictor of problems inherent in human communication as a whole (Shannon, 1949/1998)

> There is no quantum world. There is only
> an abstract quantum physical description.
> It is wrong to think that the task of physics
> is to find out how nature is. Physics
> concerns what we can say about nature
> (French and Kennedy, 1985: 305).

Most theories become a form of fact, i.e. direct application to the observable world, usually at the loss of the integrity of the concept in the process. That "there is no quantum world" only an abstraction of a quantum physical description, gives reason to the study of this phenomena and lucidly exposes Bohr's need to have everyone 'think' about the subject, not just passively accept a theory and then arbitrarily assign it to a science fiction framework from which to work in.

> But our problem is not that we do not have
> adequate concepts. What we lack is a

sufficient understanding of the unambiguous applicability of the concepts we have (French and Kennedy, 1985: 305).

It is the necessary understanding of an ideal or unambiguous application of the concept. The number of concepts is irrelevant, the quality of their communication is the point. Quality of the message sent, not quantity of the messages, is the goal for communication.

Content, in relation to context of that content, is the primary way to keep the information signal valid. New demands, because of a change in content, will change the corresponding context of that content signal. Without this change or modification, the system faces an entropy state, reducing the original quality of the information signal. The very fact that both content and context change, i.e. are variables, makes the system a developing, or evolving, process, able to grow with both language change and conceptual shifts in the observable world.

Fredrick Shiller's view of linguistic signs representing primarily the structure of intellect and not reality itself, parallels Bohr's notion that conceptual models of a quantum world are only our abstract notions of what measurement at a quantum level entails, not a perception of the world as a quantum state. Symbols reflect ideas, but such symbols are constrained by semantic and semiotic 'baggage' that interferes with the unambiguous 'meaning' of what that symbol represents. Symbols reflect the intellect not the reality of the environment of that intellect.

In Charles Morris's **Foundations of the Theory of Signs**, he notes that civilization is dependent on signs and systems of signs, and that 'the human mind is inseparable from the functioning of signs – if indeed mentality is not to be identified with such functioning.' (Innis, 1985: preface). Morris continues with 'Indeed, it does not seem fantastic to believe that the concept of a sign may prove as fundamental to the sciences of man as the concept of the atom has been for the physical sciences or the concept of the cell for the biological sciences' (Innis, 1985: preface). As

Morris states form his **Writings on the General**

Theory of Signs:

> And since it will be shown that signs are simply the objects studied by the biological and physical sciences related in certain complex functional processes, any such unification of the formal sciences on the one hand, and the social, psychological, and humanistic sciences on the other, would provide relevant material for the unification of these two sets of sciences with the physical and biological sciences (Morris, 1971: 18).

Morris continues:

> Semiotic supplies a general language applicable to any special language or sign, and so applicable to the language of science and specific signs which are used in science (Morris, 1971: 18).

An important trait of signs and symbols are there accurate use and contextualization of use as a communicative medium. W.V.O. Quine gives an idea example of this in the terms 'ambiguity' and 'vagueness'. Quine states that ambiguity differs from vagueness in that vague terms are only dubiously applicable to marginal objects, but ambiquiguous terms maybe at one true of various objects, but also false of them (Quine, 1960: 129).

Vygotsky, the Russian semiotically influenced psychologist, thought that 'words' were paradigmatic in that a general reflection upon reality is the basic characteristic of words (Innis, 1985: preface). Words play a central part in the historical growth of consciousness as a word is a microcosm of human consciousness.

The linguistic sign, as pointed out by Saussure, is a double entity, in that a word is a fundamental unit in that it functions as a linguistic sign that units not a 'thing and a 'name', but a 'concept' and a 'sound image' (Innis, 1985: 24). Saussure means by 'sign' is the indissoluble union of the two components, an association, and is not a nomenclature (Holdcroft, 1991: 48). It is not 'a naming process only or a word list' but as Rene Thom would state: 'a foundation of images from a model appears like a manifestation of the irreversible character of a universal dynamics: the model ramifies into an image isomorphic with itself' (Innis, 1985). Morris felt that the theory of signs has a two-fold relation to all other sciences in that it is both a science among the sciences and as an individual science, semiotics studies the things or properties of things in their function of serving as signs. But as every science makes use of and expresses its results in terms of signs, metascience, the science of science, must use semiotics as an 'organon' (Noth, 1995: 49).

The linguistic aspects of science can be found in a paper by the premier structural linguist of the 20th Century, Leonard Bloomfield. In his **Linguistic Aspects of Science** (1939) Bloomfield states that 'in connection with science, language is specialized in the direction of forms which successfully communicate handling responses and lend themselves to elaborate reshaping (calculation)' (Bloomfield, 1939: 55). Bloomfield concludes his paper with the statement that since mathematics and logic are verbal activities, both of these disciplines presuppose linguistics as a fundamental feature of communication (Bloomfield, 1939: 56). Bloomfield examines the semiotic value of linguistics to the physical and natural sciences and ties to the very nature of scientific thought and discourse.

BRADLEY S. TICE

Language and the Physicist

Niels Bohr's principle of complementary states that a quantum system possess properties that are equally real but mutually exclusive. Wave-particle duality is perhaps the best example of this in that during an experiment the quantum behaves either like a wave or a particle. Englert has addressed this issue of duality and finds such classical definitions of duality to be misleading at best.

> The notions of particle and wave are associated with mental pictures that are borrowed from classical (i.e. prequantum) physics. These associations are dangerous because of their obvious limitations. Therefore, "wave – particle duality", should perhaps be abandoned in favor of a more neutral term, such as

"interferometric duality" or simply "duality".
The general formulation of this concept
could read as follows: Duality-the
observations of an interference pattern and
the acquisition of which-way information
are mutually exclusive (Englert, 1996:
2154).

It has been said that Bohr was driven to invent

quantum mechanics because the stability of

matter required him to do so in that the gross

properties of the materials of the real world, their

color, conductivity, and strength, reflect the details

of their quantum states (Cahn, 1998: 57). Bohr

may also have been quantum mechanics greatest

mandarin according to Mara Beller:

While publicly abstaining from criticizing
Bohr, many of his contemporaries did not
share his peculiar insistence on the
impossibility of devising new nonclassical
concepts – an insistence that put rigid

structures on the freedom to theorize (Beller, 1998: 33).

Bohr's main point from his addressing the question of quantum theory is this: "quantum mechanics renders meaningless the question: Does light or matter consist of particles or waves? Rather, one should ask: Does light or matter behave like particles or waves?"(Pais, 1991: 22). Bohr's own contributions to quantum mechanics came after the seminal years 1925-1926 and in fact, it was when Bohr had left Heisenberg for a ski trip in Norway, that Heisenberg discovered his uncertainty relations (Pais, 1991: 22).

Early influences on Bohr's thinking were Spinoza's psychophysical parallelism and

Kierkegaard's philosophical style, rather than his ideas, but the turning point was Poul Martin Moller's book <u>Tale of a Danish Student</u> (Moore, 1966: 15). In this tale Bohr could see himself, in that the question of language and the use of language for the objective communication of experience was the very heart of the problem (Moore, 1966: 16).

Heisenberg was also interested in philosophy, but this was second to the formalizing of the quantum theory, and needs addressing in light of the many published works by Heisenberg over the years (Cassidy, 1992: 255). Cassidy notes of Heisenberg:

> Systematic philosophical positions always were of lesser importance to him [Heisenberg]. Some readers have not fully appreciated that Heisenberg's systematic philosophical pronouncements were always tailored for public consumption and were thus informed and motivated to great extent by his personal aims in addressing each particular audience (Cassidy, 1992: 255).

Most of Heisenberg's philosophical writing's were derived from one of his public addresses and that one of his earliest prescriptive addresses was a lecture to the Leipzig philosophers entitled 'Epistemological problems in modern physics' (Cassidy, 1992: 255). Heisenberg was appointed a teaching chair in Leipzig, Germany, sometime after the Copenhagen interpretation, but before Hitler's ascension into power and the subsequent disruption and discord that follow in the sciences

(Cassidy, 1992: 255). His Leipzig paper was a trial run for his influential and widely read paper 'Causal law and quantum mechanics' given at the 1930 gathering of the Vienna Circle in Konigsberg (Cassidy, 1992: 255). Heisenberg's uncertainty principle was a challenge to the notion of causality in atomic processes, and causality, or the lack of it, became his major public theme (Cassidy, 1992: 255-256).

It was Heisenberg that shifted the philosopher's task from a re-evaluation of Kant's entire epistemology to a reevaluation of the concept of causality, the assignment of a specific physical cause to each and every individual phenomenon (Cassidy, 1992: 256). Both Heisenberg and Bohr, for his complementarity principle, went far beyond the narrow quantum-mechanical applications (Cassidy, 1992: 256). To a degree Isaac Newton used philosophical representations to justify his theories of which one of the most blatant is in his Principa [Philosophiae Naturalis Principia Mathematica, 1687, English translation 1729] were he describes time as "Absolute, true, and mathematical time, of itself, and from its own nature, flows equably and without regard to anything external, and by another name is called duration" is to the modern scientist without operational meaning and as such is a 'meaningless' statement. In fact Newton did not depend on the explicit use of these concepts in his own work and was merely being philosophical

(Holton and Brush, 1985: 179-180). Heisenberg was asked, in regards to quantum theory, in 1926, a year after he had invented matrix mechanics, by Einstein 'You don't seriously believe that none but observable magnitudes go into a physical theory? To which Heisenberg replyed to Einstein 'Isn't that precisely what you have done with relativity? (Bernstein, 1982: 163).

Einstein was clear about basing theory on observable magnitudes alone, it was wrong, as the opposite occurs, in that it is the theory that describes what we can observe (Bernstein, 1982: 163). Einstein had transcended the Machian view of the world in that none of his theories published in 1905 begins with 'observed facts' but with theory (Bernstein, 1982: 164) This would upset Mach's empirically observed behavioral view of the world and it would be Mach who would deny the existence of atoms, even taunting Ludwig Boltzmann by asking him 'Have you seen one?' [in regards to 'seeing' an atom] (Bernstein, 1982: 164). Philipp Lenard (1862-1947) had contested the validity of Einstein's Theory of Relativity, some of which is contained in the following passage:

> Lenard questioned the use of imaginary experiments (Gedanken-experimente) and asked why some which challenge relativity were inadmissible. Einstein gave the argument of a theoretician (which to Lenard was totally unacceptable) that only those imaginary experiments were permissible which in principle could be carried out, even if practically they were not feasible (Beyerchen, 1977: 90).

119

Lenard stated there was two ways of forming a picture of nature. One was based on explaining equations through observations and the other based on explaining observations through equations (Beyerchen, 1977: 89). Lenard argued that Relativity violated the intuitively obvious picture of nature, to which Einstein replyed that the intuitively obvious changes over time, as physics was conceptual rather than intuitive (Beyerchen, 1977: 89).

These types of questions could be asked to this day about any scientific theory, let alone Einstein's theories, and while some of Lenard's points have 'intellectual' merit, it is Einstein's theories that have stood the test of time. Unfortunately, both Lenard, awarded the Nobel Prize in 1905 for his work on cathode rays, and Johannes Stark (1874-1957), awarded a Nobel Prize in 1919 for his work on Doppler and 'Stark' effects, were both prominent in the National

Socialist party and in the reorganization of science in Germany after the Weimer years. Beyerchen writes well about this dichotomy of how two intelligent men could be drawn into such an 'anti-intellectual' movement as the National Socialists (Beyerchen, 1977).

How language is used can also 'make' or 'break' a scientific theory as related by Goldstein:

> Bohmian mechanics was discovered in 1952 by David Bohm,…(Unfortunately, Bohm's formulation involved unnecessary complications and could not deal efficiently with spin. In particular, Bohm's invocation of the "quantum potential" made his theory seem artificial and obscured its essential structure (Goldstein, 1998a: 40).

Note that the language Bohm used in the 'coining' of the term "quantum potential" resulted in that aspect of the theory seeming 'artificial'; i.e. false, and that "obscured"; i.e. to hide, the very nature of the fundamental structure.

In the end Niels Bohr concluded that in spite of the refinement of terminology in quantum mechanics in relation to classical vocabulary that was based on empirical and theoretical considerations, all unambiguous communication about our sensory experience is ultimately based on common language that has adapted to orientation in our environment and that defines the relationship between cause and effect (Faye, 1991: 189).

Levy-Leblond's paper **Quantum Physics and Language** (1988) mentions that words are rarely, if ever, invented 'ex nihilo' in that they are borrowed from other fields or other languages and their meaning is displaced and stretched to fit the new context (Levy-Leblond, 1988: 314). Levy-Leblond makes note of the oxymoron "Quantum Mechanics" in that the quantum aspect of the word removed the mechanics aspect of the concept (Levy-Leblond, 1991: 315).

The question of wave or particles can be addressed as follows "the objective statement of the link between quantum and classical physics is that under certain specific circumstances, a quanton may approximately behave either as a classical particle or as a classical wave"(Levy-Leblond, 1991: 316). The "quanton" is a modern term, suggested by M. Bunge, that reflects the quantum concept to a quantum language and removes the weak and inaccurate classical physics view of the wavefunction/particle duality scenario (Levy-Leblond, 1991: 316, Bozic, Maric, and Vigier, 1992: 1327, and Bozic and Maric, 1995: 159).

Language is also used in mathematics as both a symbolic language and as a common form of communication of mathematical ideas. In 1900 in Paris, France at the 2nd International Congress of Mathematicians, David Hilbert (1862-1943), a

German, listed 23 problems for development for the 20th Century (Coveney and Highfield, 1995: 25-26). Hilbert's second problem was a challenge to demonstrate consistency of axioms of mathematics; i.e. Hilbert wanted mathematics to be reduced to pure logic (Coveney and Highfield, 1995: 26). Hilbert wanted a set of axioms and rules of reasoning from which could be generated all mathematical truth with no contradictions (Coveney and Highfield, 1995: 27).

This step-by-step process of procedures for carrying out operations by application of specific rules is named for an Arab scholar and are called algorithms (Coveney and Highfield, 1995: 27). It would be the 25 year old logician Kurt Godel

(1906-1978), residing in Vienna, Austria, that in a paper submitted to the journal "Monatshefte fir Mathematik und Physik" in 1930 exposed the simple fact that certain mathematical statements can neither be proved nor disproved and that there must always be statements whose truth value is undecidable (Coveney and Highfield, 1995: 28). In essence Godel showed the inevitability of finding logical paradoxes in arithmetic that are the equivalent of the statement "This sentence is false" (Coveney and Highfield, 1995: 28). While Godel's work demolished Hilbert's plan for the foundations of mathematics, it gave to the world the idea of computation and the notion of computability (Coveney and Highfield, 1995: 29). What then does Godel's undecidability problem have on physics? First, a list of precise assumptions that underlie Godel's deduction of incompleteness (Barrow, 1998: 222). Godel's theorem defines a formal system as:

1.) Finitely specified.

2.) Large enough to include arithmetic.

3.) Consistent.

How do these effect the nature of physics? In principle, incompleteness, in practice, looks very much like inadequacy in theory (Barrow, 1998: 223). New arithmetics will still be incomplete, but they always can be extended to accommodate any incompleteness. Physical theory likewise can be enlarged by adding new principles which will force all the undecidability into that part of the mathematical realm which has no physical manifestation (Barrow, 1998: 223-224, also see Rucker, 1982).

In an article by Smale in <u>The Mathematical Intelligencer</u> (1998) Smale answers a call for suggestions by mathematicians, International Mathematical Union, for problems to be addressed in the 21st Century. This was inspired by Hilbert's list of 1900 and the area of most interest from these problems is Problem #18 by Smale that states 'what are the limits of intelligence, both artificial and human?' (Smale, 1998: 13). Smale suggests a search for similarities and differences in artificial and human intelligence and that the areas of problem-solving and games theory are ideal fields of inquiry (Smale, 1998: 13). While this is a very superficial presentation of some important questions, hence the publication in the 'Reader's Digest' of mathematical journals, this is nonetheless a start in forming specific questions for the next hundred years of man's intellectual quest.

An interesting allegorical tale of numeration and the concept of infinities is addressed in Benardete's <u>Infinity: An Essay in Metaphysics</u> (1964) where he describes how the Swahilis use a fable to describe their concept of infinity[2].

---

2. The Swahili language is the most widely spoken of the Bantu languages, African, and the only one to have international status [B. Comrie (1990) <u>The</u>

Swahilis believe that the stones in their valley are literally uncountable and hence literally without number, they also believe that they are infinitely many in count (Benardete, 1964: 193). Benardete then parallels this 'proto-concept' of the Swahili's with modern man's 'standard-concept' of the intellectual world and find them similar in application, i.e. few if any of us would bother to actually count the number of stones in a valley, and that the concept of infinity, i.e. a never ending continuum of numbers, fills a similar void in our mathematical education (Bernardete, 1964: 194-195).

World's Major Languages: New York, Oxford University Press, pp. 994]. I was honored to deliver a paper "The Priority Method: African ESL" at the 2nd World Congress of African Linguistics, University of Leipzig, Leipzig, Germany, July 27-August 3, 1997.

Research

The research will consist of a Lexical-Semantic Model of Analysis that will be used to correlate a concept with a mental dictionary (lexical) and the appropriateness of that word meaning (semantics) to the conceptual meaning of the object or processes being defined.

The research will consist of the following:

1.) Definition of all terminology.

2.) Specific definition of the Lexical-Semantic Model of Analysis.

3.) Citing examples of language use in the physical sciences, especially that relating to Relativity and Quantum theories.

4.) Examples of language used in the sciences defined by the Lexical-Semantic Model of Analysis.

5.) Results of this experiment.

6.) Discussion of results.

From this the research component of this thesis will be addressed.

1.)   **Definition of All Terminology.**

The following is the precise definitions of all terminology used in this research thesis section:

Abstract: Dealing with a subject in its abstract aspects [theoretical].

Language: A systematic method of communicating ideas.

Lexical: Relating to words or the vocabulary of a language.

Physical: Having material existence.

Physics: A science that deals with matter and energy and their interactions.

Quantum: One of the very small increments or parcels into which many forms of energy are subdivided.

Science: Such knowledge concerned with the physical world and its phenomena.

Semantic: The study of linguistic meaning of morphemes; words, phrases, and sentences.

Theory: Analysis of facts in their relation to one another.

These terms will be the backbone of the language used to analyze concepts into words and meanings.

2.)   **Specific Definition of the**

Lexical-Semantic Model of Analysis.

The research will consist of a Lexical-Semantic Model of Analysis that will be used to correlate a concept with a mental dictionary (lexical) and the appropriateness of that word meaning (semantics) to the conceptual meaning of the object or processes being defined.   A flow diagram [direction from left to right] of this model is presented below [Figure 1].

Figure 1

A B C D E F G H I

F[t]

Key:_____

A: Presentation of original scientific term.

B: Valuation of present term with concept.

C: Present lexical value of term.

D: Present semantic value of term.

E: Search for better lexical/semantic value of term.

F: Propose a new definition or word, i.e. a new term for concept.

F[t]: No new definition or word.

G: Present new definition or word with supporting evidence.

H: Discussion

I: Summary

3.) **Citing Examples of Language Use in the Physical Sciences, Especially that Relate to Relativity and Quantum Theories.**

In the following four examples, three already addressed in the text of this thesis and one new idea [concept], are concepts developed around definitions that appropriately parallels the words used to describe those processes or concepts. The following are a list of those four examples:

A.) Reason Determinism

B.) Human Consciousness/Observations

C.) Quanton

D.) Hidden Variables

A.)   **Reason Determinism**

The question of 'reason determinism' as addressed by Pap (1962) is not so much 'our' knowing positions of subatomic particles but that even the unknown can still be considered a form of 'determinism', i.e. our knowledge of position is irrelevant to whether the fact that the positions of subatomic particles do exist. Our knowledge of 'preordaining' a position is absurd. Human reason is a logical abstract of measured places and times and is still a form of reason even when such measurements cannot be obtained. As determinism is still a form of human conceptual perception to the formulation of space and time to an unobserved event such that is found in the quantum world. They are not diminished by the lack of information about them. They are only less satisfying and not as complete.

B.)   Human Consciousness/Observations

Human consciousness is always a factor in observations and deductions and only plays a role in observing, and not the destiny of what is observed. We are humans not gods. The act of observation is a function of using human perceptual organs to percieve the world around us and give a conceptual model of that world through the notion of the collective use of the senses and the human brain which is, as a whole, the human mind.

C.)   Quanton

The question of wave or particles can be addressed as follows "the objective statement of the link between quantum and classical physics is that under certain specific circumstances, a quanton may approximately behave either as a classical particle or as a classical wave"(Levy-Leblond, 1991: 316). The "quanton" is a modern term, suggested by M. Bunge, that reflects the quantum concept to a quantum language and removes the weak and inaccurate classical physics view of the wavefunction/particle duality scenario (Levy-Leblond, 1991: 316, Bozic, Maric, and Vigier, 1992: 1327, and Bozic and Maric, 1995: 159).

D.)   **Hidden Variables**

David Bohm's 'Hidden Variables' (1952, 1966) is a questionable term as stated in Sharpe (1993):

> The term hidden variable can be confusing when first encountered. What makes a hidden variable hidden? How can we know and talk about it if it remains hidden? J.H. Tutsch notes that the name is inadequate because it has many interpretations. Thirty-two years after Bohm introduced his hidden variables thesis, he and Hiley said the choice of the term was unfortunate. The variables are not really in hiding; they are revealed when certain measurements are made. Popper would add that all variables start hidden. They only stop being so when the theories using them are successful (Sharpe, 1993: 22).

Bell (1987) raises a similar complaint with regards to the term 'Hidden Variables' and suggests an alternative:

> Perhaps uncontrolled variable would have been better, for these variables, by hypothesis, for the time being, cannot be manipulated at will by us (Bell, 1987: 92).

The term 'Hidden Variables' will be examined, as will the other three terms, in the Lexical-Semantic Model of Analysis in the subsequent section.

4.)   **Examples of Language Used in the Sciences Defined by the Lexical-Semantic Model of Analysis.**

The research will consist of a Lexical-Semantic Model of Analysis that will be used to correlate a concept with a mental dictionary (lexical) and the appropriateness of that word meaning (semantics) to the conceptual meaning of the object or processes being defined. A flow diagram [direction from left to right] of this model is presented below [Figure 2].

Figure 2

A B C D E F G H I

F[t]

Key:_____

A: Presentation of original scientific term.

B: Valuation of present term with concept.

C: Present lexical value of term.

D: Present semantic value of term.

E: Search for better lexical/semantic value of term.

F: Propose a new definition or word, i.e. a new term for concept.

F[t]: No new definition or word.

G: Present new definition or word with supporting evidence.

H: Discussion

I: Summary

The following four terms will be defined by the Lexical-Semantic Model of Analysis.

A.) Reason Determinism

B.) Human Consciousness/Observations

C.) Quanton

D.) Hidden Variables

Experiment #1

Term: Reason Determinism

Lexical-Semantic Model of Analysis

A: Presentation of original scientific term.

Reason Determinism: A logical formalism of determinist thought to quantum environments.

The notion that conscious activity pre-ordains the physical world rather than observes it in an embedded, yet non-deterministic manner.

B: Valuation of present term with concept.

A philosophical question that is open to interpretation. Weak structure and poor execution of word to concept. Reason is a logical function that has modern roots dating back to the ancient Greeks and has the emphasis of being 'empirically' based and 'scientifically' rational as a philosophy. This gives a 'false' sense of truth value to the notion being presented in that 'determinism' is hence forth a 'reasonable' or 'reasoned', i.e. a logical truth, expression.

C: Present lexical value of term.

Lexical value is strong. The terminology is viable as a word combination. Both words found in the English language vocabulary.

D: Present semantic value of term.

Weak value of semantic definition of word value to philosophical value. Easy to undermine core concept. 'Selling' a concept rather than developing one, a rather underhanded method of gaining credability.

E: Search for better lexical/semantic value of term.

The question of 'reason determinism' is not so much 'our' knowing positions of subatomic particles but that even the unknown can still be considered a form of 'determinism', i.e. our knowledge of position is irrelevant to whether the fact that the positions of subatomic particles do exist.

F: Propose a new definition or word, i.e. a new term or concept.

New Definition: Reason Determinism does not influence events in the world, it just reflects impressions of that world. Also semantic value of words in question.

F[t]: No new definition or word.

G: Present new definition or word with supporting evidence.

New Definition: Reason Determinism does not influence events in the world, it just reflects impressions of that world. Perhaps a better term would be 'Predictive Determinism' in that determinism is predictive, but not a logical or reasoned event.

H: Discussion

Our knowledge of events does not predict those events, just acknowledges that they occur. The conceptual and semantic value of this term is in question. A new term would be more accurate, but the question of a pre-ordaning view of the universe is, again, a very narrow interpretation and of questionable support.

I: Summary

A new definition to the word that reflects a constraint to human actions to perceived world events. Proposed 'Predictive Determinism' as an alternative to 'Reason Determinism'.

Experiment #2

Term: Human Consciousness/Observations

Lexical-Semantic Model of Analysis

A: Presentation of original scientific term.

Human consciousness with the emphasis on interacting on and preordaining observations of the physical world.

B: Valuation of present term with concept.

Weak philosophical valuation of concept to semantic values of terms. Human consciousness is not necessarily a 'pre-ordaining' factor to human consciousness. Also observations are just those perceptual valuations of the world taken in by sense organs and the human brain to form a view of the physical world around us.

C: Present lexical value of term.

Lexical value is strong. Word formation valid. Both words found in the English language vocabulary.

D: Present semantic value of term.

Poor support for human influence on observables. Human interaction is not necessarily human control of observables. These "pre-ordaining' values effect both the concept of human influence and the act of observation. Human influence is just local and embedded in the environment and is the manifestation of the value of the 'conceptual' interaction of man to his environment. Observables are those qualities that are found in the human act of observing. It is just the 'behavior' of observing and not the 'interaction' of that behavior that is defining because all observations are embedded into the environment and hence the act of observing.

E: Search for better lexical/semantic value of term.

Human consciousness is always a factor in observations and deductions and only plays a role in observing, and not the destiny of what is observed

F: Propose a new definition or word, i.e. a new term for concept.

New definition would reduce the effect of human interaction to just observing and not influencing observed world events.

F[t]: No new definition or word.

G: Present new definition or word with supporting evidence.

Human consciousness is a passive influence on the object world. Perhaps coining the term as

'Interpretive Observations' or 'Assessed Observations' would be a more accurate choice in that the desired operative -pre-ordaining states – is an interpretive view of observations and is the 'assessed' form of those observations.

H: Discussion

A new definition is necessary to reduce the cause and effect model of human perception and observed events. The term 'observables' is not limited to just one interpretation and is relatively free from notions outside of 'that which is observable'.

I: Summary

Change of definition of terms to reduce human influences on observables. 'Observables' is a less prejudgical term and is a more accurate term in defining the act of observing.

Experiment #3

Term: Quanton

Lexical-Semantic Model of Analysis

A: Presentation of original scientific term.

A quanton may approximately behave either as a classical particle or as a classical wave. A hybride term that is both a classical particle and a classical wave and reflects the duality of the nature of the quantum world.

157

B: Valuation of present term with concept.

Good valuation in word to concept. Represents the dual nature of the experimental aspects of the quantum world.

C: Present lexical value of term.

Strong lexical value of word. Valid word hybridization from the English language.

D: Present semantic value of term.

Strong semantic value of word. The hybride word justifies the dual nature of the quantum world by inference to experimental results of particle and wave duality.

E: Search for better lexical/semantic value of term.

None.

F: Propose a new definition or word, i.e. a new term for concept.

F[t]: No new definition or word.

No new definition or word needed.

G: Present new definition or word with supporting evidence.

H: Discussion

The fact that the term quanton represents the quantum events as approximately behaving either as a classical particle or as a classical wave makes the lexical and semantic values strong for this word.

I: Summary

No change to definition or word.

Experiment #4

Term: Hidden Variables

Lexical-Semantic Model of Analysis

A: Presentation of original scientific term.

Hidden variables are hiding until 'discovered' by new experiments. They exist regardless of physical proof. Pre-conceived notion of shadow elements in the quantum world.

B: Valuation of present term with concept.

Weak lexical and semantic values to concept to word.

C: Present lexical value of term.

Lexial value of term valid. Both words found in English language vocabulary.

D: Present semantic value of term.

The fact that the variables are hidden presupposes that they exist in the first place. Hidden variables is a confusing term because it presupposes something that has not been found to exist. "Hidden' implies that it exists and has not been found yet by experiment. Also 'variables' implies that it interacts with the quantum environment as a variable, a condition of change in state.

E: Search for better lexical/semantic value of term.

The variables are not really in hiding; they are revealed when certain measurements are made that detect them. This needs to be addressed in defining a new term.

F: Propose a new definition or word, i.e. a new term for concept.

Perhaps 'uncontrolled' variable would have been better, for these variables, by hypothesis, for the time being, cannot be manipulated at will by us.

F[t]: No new definition or word.

G: Present new definition or word with supporting evidence.

Uncontrolled Variable would be a more scientific term than hidden variables. The semantic value of the term would parallel the actual nature of such a variable.

H: Discussion

The term hidden variable can be confusing when first encountered. What makes a hidden variable hidden? How can we know and talk about it if it remains hidden? A definition to the term is necessary to remove poor semantic value and lexical value of the original word.

I: Summary

Proposed 'Uncontrolled Variables' for Hidden Variables.

5.)   **Results of this Experiment.**

The following are the results from the four experiments using the Lexical-Semantic Model of Analysis to analyze the following four terms:

A.) Reason Determinism

B.) Human Consciousness/Observations

C.) Quanton

D.) Hidden Variables

Experiment #1

A.) Reason Determinism

Discussion:

Our knowledge of events does not predict those events, just acknowledges that they occur.

Summary:

A new definition to the word that reflects a constraint to human actions to perceived world events.

Experiment #2

B.) Human Consciousness/Observations

Discussion:

A new definition is necessary to reduce the cause and effect model of human perception and observed events. Perhaps coining the term as 'Interpretive Observations' or 'Assessed Observations' would benefit this concept.

Summary:

Change of definition of terms to reduce human influences on observables.

Experiment #3

C.) Quanton

Discussion:

The fact that the term quanton can approximately behave either as a classical particle or as a classical wave makes the lexical and semantic values strong for this word.

Summary:

No change to definition or word.

Experiment #4

D.) Hidden Variables

Discussion:

The term hidden variable can be confusing when first encountered. What makes a hidden variable hidden? How can we know and talk about it if it remains hidden? A definition to the term is necessary to remove poor semantic value and lexical value of the original word.

Summary:

Proposed 'Uncontrolled Variables' for Hidden Variables.

6.) **Discussion of Results.**

The Lexical-Semantic Model of Analysis was used to evaluate the lexical and semantic strength of four terms that were measured for lexical and semantic values that then addressed the issue of improving upon these values and then proposing a new definition or word to meet the new valuations if necessary.

Of interest is that some words did not need new 'coinage' or new definitions, but the words that did change reflected the conceptual needs for language to parallel conceptual models of human valuations of world events. This is the essence of the Lexical-Semantic Model of Analysis as it reflects the needs of a concept to be accurately defined by a word or words. In some respects the Lexical-Semantic Model of Analysis tests the value of a concept to word measure that is necessary for a functional use of the word as a verbal component of thought reflecting the meaning measure of the concept desired.

Summary

Language is a manner to communicate ideas. The very nature of language is based on the transfer of one symbolic representation from one person to another. Because a common core network of symbols to concepts is the very nature of effective transmittal of ideas, the language used to express those concepts must accurately convey those ideas from the speaker to the listener.

Language change is the norm, not an exception, and such change is facilitory to the main goal of language: communication. When words are used to convey concepts that are inherent to that word, but undesired for the new use of that word, i.e. a new concept, then the language fails to convey the total accuracy desired of the use of that word. With quantum physics the use of classical terminology has made the use of words tied to the old physics, classical or Newtonian physics, problematic for the new fields, i.e quantum physics.

The question to ask is what is to be done about this fossilization of language meaning in the great flow of language change? New terms can help, but more often than not, clear explanations of terminology would also go far in reducing abstractions from the desired meaning of words. Words used in a specific context are the very nature of ideal communication, in science, and all other areas to be communicated. Ideas must be developed, expressed clearly, and received in an ideal state of transmission, i.e. 100% efficiency, to be be communicated effectively.

In conclusion, language change is the nature of language, such a evolutionary aspect to a human process must be accounted for in both the manner of language, but also in the type of ideas that are expressed by that language. Language and science are in flux. Keeping that in mind, when using language to describe ideas in science, is the crux of language use and the bases for the philosophy of science.

As stated in the introduction chapter to this dissertation the following question was presented to be addressed using relativity theory and quantum theory as subfields within the physical sciences. As stated from that chapter:

> When we discuss or describe a process or concept, we are always utilizing existing terminology to describe that process or concept, even though such a process or concept may not 'fit' such existing descriptions. This then is the major task facing the use of language in the field of science.

From this a review of authors of varied backgrounds is addressed to give a literary pool of though on science that is then focused on the matters of language and science. Examples were given from Heisenberg, Bohr, Bohm and

BRADLEY S. TICE

Schrodinger, all world renouned physicists, in the use of language verses both the observable world, i.e. experimental, and the unobservable world, i.e. the quantum world, and how this shapes concepts and language used to relate to those concepts.

The research section of this dissertation presented the Lexical – Semantic Model of Analysis that resulted in the need to test a word to a concept to find a measure of value to the lexical and semantic traits of a word as it relates to the concept being expressed.

What then is the conclusion, or answer, to this question of language and science? As the main just of this work has focused on the 'ambiguous' nature of language and the nature of language change, the answer is to focus on a clear and developing language of terminology that, while minimizes ambiguity, also realizes the natural stress and change of terminology to fit all domains of language representation. What exactly does this mean? One is the development and expansion of word definitions. Another is to 'coin' new expressions that are 'absolutely necessary' for a new conceptual process or idea. One hopes to stay away from the 'word mills' affecting other disciplines, like the notorious social sciences, and that terminology will avoid fashionable excesses, such as the use of of C.E.

for Current Era rather than A.D., Anno Domini, to designate a specific timeline.

From this a general theory of language and science can be modelled:

The use of language must reflect the observable nature of a scientific observation and that the abstract nature of a concept must exist within the domain of the rules of grammar that form that of language. Translations of this language must strive to accurately reflect the true nature of the semantics present in the original language and that additions or explanations maybe necessary to aid in this linquistic realization. Language is the communication mode of scientific thinking and expression. Language must reflect the accuracy of the semantics found in the ideas and expressions found in science. Language does not create in the sciences. It just reflects that which is created by science. Language is utilization. Science is creation of ideas. Language merely communicates those ideas.

This then is the sum total of the answer to the question posed at the beginning of this dissertation. The question of language as related to the expression of science is really the flexibility of that language to model the ideas and concepts envoked from the sciences.

BRADLEY S. TICE

References

Afriat, A. and Selleri, F. (1999) The Einstein, Podosley, and Rosen Paradox in Atomic, Nuclear and Particle Physics. New York: Plenum Press.

Adler, S. L. (1995) Quaternionic Quatum Mechanics and Quantum Fields. New York: Oxford University Press.

Agazzi, E. (1998) "Waves, particles, and complementarity". In The Nature of Quantum Paradoxes. Edited by Tarrozzi, G., and M. Alwyn van der Merwe. Dordrecht: Kluwer Academic Publishers.

Alonso, M. and Valk, H. (1973) Quantum Mechanics: Principles and Applications. Menlo Park: Addison-Wesley Publishing Company.

Barrow, J.D. (1998) Impossibility: The Limits of Science and the Science of Limits. Oxford: Oxford University Press.

Barut, A.O., Feranchuk, I.D., Shnir, Y.M., Tounilichik, L.M. (1995) Quantum Systems: New Trends and Methods. Singapore: World Scientific.

Bauer, H.H. (1992) <u>Scientific Literacy and the Myth of the Scientific Method</u>. Urbana: University of Illinois Press.

Bell, J.S. (1987) <u>Speakable and Unspeakable in Quantum Mechanics</u>. Cambridge: Cambridge University Press.

Beller, M. (1998) "The sokal hoax: At whom are we laughing". <u>Physics Today</u>, Volume 51, number 9, September 1998, pp. 29-34.

Benardete, J.A. (1964) <u>Infinity: An Essay in Metaphysics.</u> Oxford: Clarendon Press.

Bergstein, T. (1972) <u>Quantum Physics and Ordinary Language.</u> Humanities Press.

Berlin, B. and Kay, P. (1969) <u>Basic Color Terms</u>. Berkeley: University of California Press.

Bernstein, J. (1982) <u>Science Observed</u>. New York: Basic Books, Inc., Publishers.

Bernstein, J. (1991) <u>Quantum Profiles.</u> Princeton: Princeton University Press.

Bethe, H. and Jackiw, R. W. (1964) <u>Intermediate Quantum Mechanics</u>. New York: W.A. Benjamin, Inc.

Beyerchen, A.D. (1977) <u>Scientists under Hitler:</u> <u>Politics and the Physics Community in the</u> <u>Third Reich</u>. New Haven: Yale University Press.

Bitsakis, E. (1985) "Is it possible to save causality and locality in quantum mechanics?". In Tarozzi, G. and Van der Merwe, A. <u>Open</u> <u>Questions in Quantum Physics</u>. Dordrecht: D. Reidel Publishing Company, pages 63-73.

Blinder, S. M. (1974) <u>Foundations of Quantum</u> <u>Dynamics.</u> New York: Academic Press.

Black, T.D., Nieto, M.M., Pilloff, H.S., Scully, M.O., Sinclair, R.M. (1992) <u>Foundations of</u> <u>Quantum Mechanics</u>. Singapore: World Scientific.

Bloomfield, L. (1939) "Linguistic aspects of science". <u>International Encyclopedia of</u> <u>Unified Science</u>, Volume 1, Number 4. Chicago: The University of Chicago.

Bockhoff, F.J. (1969) <u>Elements of Quantum</u> <u>Theory</u>. Menlo Park: Addison Wesley Publishing.

Bohm, A. (1979) <u>Quantum Mechanics.</u> New York: Springer-Verlag.

Bohm, D. (1951) <u>Quantum Theory.</u> Englewood Cliffs: Prentice-Hall, Inc.

Boorstin, D.J. (1983) <u>The Discoverers</u>. New York: Random House.

Bonifacio, R. (1999) <u>Mysteries, Puzzles, and Quantum Paradoxes in Quantum Mechanics.</u> Woodbury: AIP Conference Proceedings 461.

Borowitz, S. (1967) <u>Fundamentals of Quantum Mechanics</u>. New York: W.A. Benjamin, Inc.

Bozic, M. and Maric, Z. (1995) "Compatible statistical interpretation of a wave packet". <u>Foundations of Physics</u>, Volume 25, Number 1, January 1995.

Bozic, M., Maric, Z., and Vigier, J.P. (1992) "De broglian probabilities in the double-slit experiment". <u>Foundations of Physics</u>, Volume 22, Number 10, October 1992.

Brandt, S. and Dahmen, H.D. (1995) <u>The Picture Book of Quantum Mechanics</u>. New York: Springer-Verlag.

Broglie, L. de (1953) <u>The Revolution in Physics.</u> New York: The Noonday Press.

Brown, R.W. (1956) <u>Composition of Scientific Words</u>. Washington: Published by Author.

Bub, J. (1997) Interpreting the Quantum World. Cambridge: Cambridge University Press.

Bud, J. (1993) "Measurement and Objectivity in Quantum Mechanics". In Doebner, H.D., Scherer, W., Schroeck, F. (1993) Classical and Quantum Systems: Foundations and Symmetries. Singapore: World Scientific, pages 9-18.

Bunge, M. (1967) Quantum Theory and Reality. New York: Springer-Verlag.

Cahn, R.N. (1998) "Particle physics and our everyday world: A Reply". Physics Today, Volume 51, Number 11, November 1998, pp. 57-58.

Carroll, J.B. (1956) Language, Thought, and Reality: Selected Writings of Benjamin L. Whorf. Cambridge: MIT Press.

Cassels, J.M. (1970) Basic Quantum Mechanics. London: McGraw-Hill.

Cassidy, D.C. (1992) Uncertainty: The Life and Science of Werner Heisenberg. New York: W. H. Freeman and Company.

Cohen-Tannoudji, C., Diu, B., and Laloe, F. (1977) <u>Quantum Mechanics</u>. New York: John Wiley & Sons.

Coveney, P. and Highfield, R. (1995) <u>Frontiers of Complexity</u>. New York: Fawcett Columbine.

Cropper, W.H. (1970) <u>The Quantum Physicists.</u> New York: Oxford University Press.

Davis, P.C.W. and Betts, D.S. (1994) <u>Quantum Mechanics</u>. London: Chapman & Hall.

Davydov, A.S. (1985) <u>Quantum Mechanics.</u> Oxford: Pergamon Press.

D'Espagnat, B. (1971) <u>Conceptual Foundations of Quantum Mechanics</u>. Menlo Park: W.A. Benjamin, Inc.

Dirac, P.A.M. (1930) <u>The Principles of Quantum Mechanics</u>. Oxford: Clarendon Press.

Doebner, H.D., Scherer, W., Schroeck, F. (1993) <u>Classical and Quantum Systems: Foundations and Symmetries.</u> Singapore: World Scientific.

Duffey, G. H. (1992) <u>Quantum States and Processes</u>. Englewood Cliffs: Prentice-Hall Publishers.

The Economist (1999) "Parallel universes: A world apart". The Economist, May 22, 1999, pp. 91.

Eisenbud, L. (1971) The Conceptual Foundations of Quantum Mechanics. New York: Van Nostrand Reinhold Company.

Englert, B.G. (1996) "Fringe visibility and which-way information: An inequality. Physical Review Letters Volume 77, Number 11, September 9, 1996, pages 2154-2157.

Faye, J. (1991) Niels Bohr: His Heritage and Legacy. Dordrecht: Kluwer Academic Publishers.

Fong, P. (1962) Elementary Quantum Mechanics. London: Addison-Wesley Publishing Company.

French, A.P. and Kennedy, P.J. (1985) Niels Bohr: A Centenary Volume. Cambridge: Harvard University Press.

Furth, R.H. (1970) Fundamental Principles of Modern Theoretical Physics. Oxford: Pergamon Press.

Gamow, G. (1947) One, Two, Three...Infinity. New York: Viking Press.

Gamow, G. (1966) <u>Thirty Years That Shook Physics</u>. New York: Doubleday & Company.

Gleick, J. (1987) <u>Chaos.</u> New York: Viking Press.

Goethe, J.W. (1967) <u>Goethe's Theory of Colours.</u> London: Frank Cass & Co. Ltd.

Goldstein, S. (1998a) "Quantum theory without observers-Part one". <u>Physics Today</u>, Volume 51, Number 3, March 1998, pages 42-46.

Goldstein, S. (1998b) "Quantum theory without observers-Part two". <u>Physics Today</u>, Volume 51, Number 4, April 1998, pages 38-42.

Gottfried, K. (1966) <u>Quantum Mechanics.</u> New York: W.A. Benjamin, Inc.

Gribbin, J. (1984) <u>In Search of Schrodinger's Cat.</u> New York: Bantam Books.

Gribbin, J. (1995) <u>Schrodinger's Kittens and the Search for Reality</u>. Boston: Little, Brown and Company.

Griffiths, R.B. (1994) "A consistent history approach to the logic of quantum mechanics". In Laurikainen, K.V., Montonen, C., and Sunnarborg, K. <u>Symposium on the Foundations of Modern Physics</u> 1994. Gig-sur

Yvette, France: Editions Frontiers, pages 85-97.

Griffiths, R.B. and Omnes, R. (1999) "Consistent histories and quantum measurements". Physics Today, August 1999, Volume 52, Number 8, Part 1, pp. 26-31.

Guillemin, V. (1968) The Story of Quantum Mechanics. New York: Charle's Scribner's Sons.

Hartkamper, A. and Neumann, H. (1974) Foundations of Quantum Mechanics and Ordered Linear Spaces. Berlin: Springer-Verlag.

Haroche, S. (1998) "Entanglement, decoherence and the quantum/classical boundary". Physics Today, Volume 51, Number 7, July 1998, pp. 36-42.

Heisenberg, W. (1983) Tradition in Science. New York: The Seabury Press.

Heisenberg, W. (1983a) Encounters with Einstein. Princeton: Princeton University Press.

Heisenberg, W. (1930) The Physical Principles of the Quantum Theory. Chicago: Dover Publications, Inc.

Hendry, J. (1984) <u>The Creation of Quantum Mechanics and the Bohr-Pauli Dialogue</u>. Dordrecht: D. Reidel Publishing Company.

Hilgevoord, J. and Uffink, J. (1990) "A new view of uncertainty principle". In <u>Sixty Two Years of Uncertainty</u>. Edited by Miller, A.I. New York: Plenum Press, pages 121-137.

Holton, G. (1985) <u>Introduction to Concepts and Theories in Physical Science</u>. Princeton: Princeton University Press.

Hund, F., and Reece, G. (1967) <u>The History of Quantum Theory</u>. London: Harrap.

Innis, R.E. (1985) <u>Semiotics: An Introductory Anthology</u>. Bloomington: Indiana University Press.

Jammer, M. (1974) <u>The Philosophy of Quantum Mechanics.</u> New York: John Wiley & Sons.

Jauch, J.M. (1968) <u>Foundations of Quantum Mechanics.</u> Menlo Park: Addison-Wesley Publishing Company.

Jeans, J. (1958) <u>Physics and Philosophy.</u> Ann Arbor: The University of Michigan Press.

Kemble, E.C. (1937) The Fundamental Principles of Quantum Mechanics. New York: McGraw-Hill Book Company, Inc.

Kieffer, B. (1986) The Storm and Stress of Language: Linguistic Catastrophe in the Early Works of Goethe, Lenz, Klinger, and Schiller. University Park: The Pennsylvania State University Press.

Klauder, J.R. (1999) "Metrical quantization". In Quantum Future Proceedings of the Xth Born Symposium, Przesreka, Poland, September 24-27, 1997. Berlin: Springer-Verlag.

Kochen, S. (1985) "A new interpretation of quantum mechanics". In Laht, P., and Mittelstaedt, P. Symposium on the Foundations of Modern Physics 1985. Singapore: World Scientific, pages 151-169.

Kramers, H.A. (1958) Quantum Mechanics. Amsterdam: North-Holland Publishing Company.

Kursunoglu, B. (1962) Modern Quantum Theory. San Francisco: W.H. Freeman and Company.

Kwek, K.H. (1998) Frontiers in Quantum Physics. Singapore: Springer-Verlag.

Landau, L.D. and Lifshitz, E. M. (1958) <u>Quantum Mechanics.</u> London: Pergamon Press.

Leibfried, D., Pfau, T., and Monroe, C. (1998) "Shadows and mirrors: Reconstructing quantum states of atom motion". <u>Physics Today</u>, Volume 51, Number 4, April 1998, pages 22-28.

Levy-Leblond, J.M. (1988) "Quantum physics and language" in <u>Matter Wave Interferometry</u> edited by Badurek, G., Rauch, H., and Zeilinger, A. Amsterdam: North-Holland.

Lipkin, H.J. (1973) <u>Quantum Mechanics.</u> Amsterdam: North-Holland Publishing Company.

Lofgren, L. (1993) "Linguistic realism and issues in quantum philosophy". In Laurikainen, K.V., and Montonen, C. <u>Symposia on the Foundations of Modern Physic</u>s. Singapore: World Scientific, pages 297-318.

Mandl, F. (1957) <u>Quantum Mechanics. </u> London: Butterworths.

Mann, A. and Revzen, M. (1996) <u>The Dilemma of Einstein, Podolsky, and Rosen-60 Years Later</u>. Philadelphia: Institute of Physics Publishing.

Marlow, A.R. (1978) <u>Mathematical Foundations of Quantum Theory.</u> New York: Academic Press.

Marr, D. (1982) <u>Vision.</u> San Francisco: W.H. Freeman.

Maxwell, N. (1995) "A philosopher struggles to understand quantum theory: Particle creation and wave packet reduction". Ferrero, M. and Van der Meruve, A. (editors) <u>Fundamental Problems in Quantum Physics</u>. Dordrecht: Kluwer Academic Publishers, pages 205-214.

Merzbacher, E. (1961) <u>Quantum Mechanics.</u> New York: John Wiley & Sons, Inc.

Messiah, A. (1963) <u>Quantum Mechanics.</u> New York: Interscience Publishers.

Moore, R. (1966) <u>Niels Bohr</u>. New York: Alfred A. Knopf.

Morris, C. (1971) <u>Writings on the General Theory of Signs</u>. The Hague: Mouton.

Noth, W. (1995) <u>Handbook of Semiotics.</u> Bloomington: Indiana University Press.

Olkhovsky, V.S. (1998) On time as a quantum-physical observable quantity". In Bonifacio, R. (1999) <u>Mysteries, Puzzles, and Quantum</u>

Paradoxes in Quantum Mechanics. Woodbury: AIP Conference Proceedings 461.

Pais, A. (1991) Niels Bohr's Times, In Physics, Philosophy, and Polity. Oxford: Clarendon Press.

Pap, A. (1962) Philosophy of Science. New York: Free Press of Glencoe.

Park, D. (1964) Introduction to the Quantum Theory. New York: McGraw-Hill Book Company.

Peat, F.D. (1997) Infinite Potential: The Life and Times of David Bohm. Reading, Mass: Addison-Wesley.

Perkowitz, S. (1996) Empire of Light. New York: Henry Holt and Company.

Planck, M. (1932) Where is Science Going. New York: W.W. Norton & Company, Inc.

Popper, K. (1956) The Open Universe. Totowa, New Jersey: Rowman and Littlefield.

Powell, J. L. and Crasemann, B. (1961) Quantum Mechanics. London: Addison-Wesley Publishing Company, Inc.

Quine, W.V.O. (1960) Word and Object. Cambridge: The MIT Press.

Roman, P. (1960) Theory of Elementary Particles. New York: Interscience Publishers.

Ruark, A.E. and Urey, H.C. (1930) Atoms, Molecules and Quanta. New York: McGraw-Hill Book Company.

Rucker, R. (1982) Infinity and the Mind. Birkhauser. Boston.

Saussure, F. (1916) Course in General Linguistics. New York: The Philosophical Library.

Schommers, W. (1995) Symbols, Pictures and Quantum Reality: On the Theoretical Foundations of the Physical Universe. London: World Scientific.

Shannon, C.E. (1949/1998) A Mathematical Theory of Communication. 50th Anniversary Edition. Printed for the 1998 I.E.E.E. International Symposium on Information Theory, MIT, Cambridge, MA, August 16-21, 1998.

Shannon, C.E. and Weaver, W. (1949) The Mathematical Theory of Communication. Urbana: University of Illinois Press.

Sharpe, K.J. (1993) <u>David Bohm's World.</u> London: Bucknell University Press.

Slater, J. C. (1960) <u>Quantum Theory of Atomic Structure.</u> New York: McGraw-Hill Book Company, Inc.

Smale, S. (1998) "Mathematical Problems for the Next Century". <u>The Mathematical Intelligencer</u>, Volume 20, Number 2, Spring 1998, pp. 7-15.

Sopka, K.R. (1980) <u>Quantum Physics in America 1920-1935.</u> New York: Arno Press.

Spielberg, N. and Anderson, B.D. (1987) <u>Seven Ideas that Shook the Universe</u>. New York: John Wiley & Sons, Inc.

Sucher, J. (1995) "The concept of potential in quantum field theory". In Barut, A.O., Feranchuk, I.D., Shnir, Y.M., Tounilichik, L.M. (1995) <u>Quantum Systems: New Trends and Methods</u>. Singapore: World Scientific.

Swartz, N. (1985) <u>The Concept of Physical Law</u>. Cambridge: Cambridge University Press.

Tice, B. (1997a) "The Turing Machine: A Question of Linguistics" Paper presented at the 78th annual Meeting of the Pacific Division of the AAAS Corvallas, Oregon June 22-26, 1997.

Tice, B. (1997b) "The Semiotics of Science" Paper presented at the 6th Congress of the IASS-AIS Conference July 13-18, 1997 Guadalajara, Mexico.

Tice, B. (1997c) "The Chemical Bond: A Misnomer?" Paper presented at the 214th American Chemical Society Meeting September 7-11, 1997 Las Vegas, Nevada.

Tice, B. (1997d) "The Veil of Valence" Poster presented at the 214th American Chemical Society Meeting September 7-11, 1997 Las Vegas, Nevada.

Tice, B. (1997e) "A History of the Chemical Bond" Paper/Poster presented at the 214th American Chemical Society Meeting September 7-11, 1997 Las Vegas, Nevada.

Tomonaga, S.I. (1962) Quantum Mechanics. New York: Interscience Publishers.

Tsipis, C.A., Popov, V.S., Herschbach, D.R., and Avery, J.S. (1996) New Methods in Quantum Theory. Dordrecht: Kluwer Academic Publishers.

Urigu, R. "On the physical meaning of the shannon information entropy". In Busch, P., Lahiti, P., and Mittelstaedt, P. Symposium on the Foundations of Modern Physics. Singapore: World Scientific, pages 398-405.

Vygotsky, L. (1962) Thought and Language. Cambridge: MIT Press.

Waerden, van der B.L. (1967) Sources of Quantum Mechanics. Amsterdam: North-Holland Publishing Company.

Watanabe, S. (1983) "Is reductionism tenable within physics?" in Proceedings of the International Symposium on the Foundations of Quantum Mechanics. Physical Society of Japan, pages 261-264.

Wentzel, G. (1949) Quantum Theory and Fields. New York: Interscience Publishers, Inc.

Wesley, J. P. (1996) Classical Quantum Theory. Blumberg: Benjamin Wesley Publisher.

West, T.G. (1997) In the Mind's Eye. Amherst, New York: Prometheus Books.

Wieder, S. (1973) The Foundations of Quantum Theory. New York: Academic Press.

Yam, P. (1997) "Bringing schrodinger's cat to life". Scientific American, June 1997, Volume 276, Number 6, pp. 124-129.

Ziman, J.M. (1969) <u>Elements of Advanced Quantum Theory.</u> Cambridge: Cambridge University Press.

Ziock, K. (1969) <u>Basic Quantum Mechanics.</u> New York: John Wiley & Sons, Inc.

BRADLEY S. TICE

Notes

1.) I have tried to get copies of all papers cited and they are listed in the appendices of this dissertation. Perhaps in the near future, 'electronic' versions of thesis and dissertations will be able to electronically link all cited matter into a unified whole for the rapid study of citations and noted materials.

2.) Such a language paradigm is used by Levy-Leblond in his concept of 'Quantics' rather than the oxymoron "Quantum Mechanics". Words are already loaded with both a semantic and semiotic level and reflect both current and past associations and definitions.

When we apply old terms to new concepts and processes, the old definitions become a part of the new concept, and unless they parallel the new concepts description, they will interfere with the accuracy of that action or process being described. See Levy-Lebond, J.M. "Quantum physics and language" in Badurek, G., Rauch, H., and Zeilinger, A. (editors) **Matter Wave Interferometry**, Amsterdam: North-Holland.

Appendix

BRADLEY S. TICE

Appendix A

BRADLEY S. TICE

THOUGHT, FUNCTION, AND FORM:
THE LANGUAGE OF PHYSICS

PACIFIC DIVISION

AMERICAN ASSOCIATION FOR THE

ADVANCEMENT OF SCIENCE

ANNUAL MEETING

OREGON STATE UNIVERSITY

JUNE 22-26, 1997

TITLE: THE TURING MACHINE: A QUESTION

OF LINGUISTICS?

PRESENTER: BRADLEY S. TICE, ADVANCED

HUMAN DESIGN CUPERTINO, CALIFORNIA

ABSTRACT

In examining the 'imitation game' posed by Alan
Turing in the context of 'can machines think?' is rather

a question of linguistics than that of computer science[1].

Although what constitutes the theory of machine language, formal languages, is not like human, or natural languages, a need for a similar core raises the question that the Turing Machine is, in fact, a linguistic or language machine rather than a computer.

Introduction

In describing the 'imitation game' Alan Turing tries to answer the question "Can a Machine Think?". In the imitation game there are three people, (A), a woman, (B), a man and (C), an interrogator, of either gender. The object of the game is for the interrogator to determine which of the two are a man or a woman.

---

[1] Turing, A.M. "Can a Machine Think?" in James R. Newman's <u>The World of Mathematics</u> (New York: Simon and Schuster, 1956).

The answers are typewritten to disguise tone quality. Turing then switches the woman, (A), with a machine and asks 'what would happen to the game?'.

Discussion

The criteria for Artificial Intelligence is also the criteria for human communication. The real question is not so much 'who' or 'what' is communicating but rather 'how' this communicating is done. This falls into the field of language use and linguistics and this can be demonstrated by a simple revision of the 'imitation game'. In the Imitation Game, Alan Turing sets up the criteria for 'what is human' and 'what is a machine' in defining an 'intelligent machine'.

When he substitutes a machine for a women in (A) he is asking the question 'Is (A) human or not?". In essence, it is the use of language as a medium for discerning whether (A) is human or not. Non-human responses are those that are conceptually, socially, grammatically and culturally alien to (C)'s, the interrogator, speaker environment. Errors in communication can also result if we modify the human in (A). Replace (A) with a human that is deaf, dumb and blind. Such perceptual limits would have a serious effect on not only the types of output but the very output itself. A common example in everyday life could be used to replace (A) by including an individual with a language unknown to (C), the interrogator, as this would also function as a severe constraint to the output and input of both (A) and (C).

In other words, the imitation game is based on a parochial standard of monolingual and ethnocentric cultural and linguistic norms than on an actual criteria of machine verses human interaction. Such a question should be raised because we still use the 'imitation game' to define human and artificial intelligence norms when in fact we are only defining linguistic and cultural ones.

Like the question of a computer 'beating' a chess champion, such as IBM's 'Big Blue' and World Chess Champion Gary Kasparov, that had the media excited as a 'machine that beat a human'. The fact that a man made and programmed machine that has 100% memory recall of all the strategies and games ever played, against the variable stamina and intellectual peaks of a human chess champion, is not so much machine verses man, but rather human ingenuity, the machine, verses human stamina, the human chess champion.

Also chess is a very specialized form of intelligence, on par with music and mathematics; the only fields with the phenomena of 'child prodigies', and does not reflect the totality of the human intellectual capacity or traits. Chess is also the domain of eccentrics, such as Bobby Fischer, and madmen, Alexander Alekhine, as well as intellectuals, and is not really the best measure of a balanced human intellect.

This claim can be leveled at the 'imitation game' as it only reflects linguistic and cultural norms and not the more expansive traits usually associated with the human condition.

Conclusion

What is needed is a new criteria to define artificial intelligence, perhaps moving away from the 'specialized' human traits to those that are more akin to the 'behavior' of machines. Like teaching primates human sign language, we often fail to realize that human behavior and acquisition skills are specifically oriented to human beings. Other creatures and machines have their own traits and behaviors that may or may not parallel our own. In the end the 'imitation game' is just a question of language, not a question of machine intelligence.

Appendix B

THOUGHT, FUNCTION, AND FORM:
THE LANGUAGE OF PHYSICS

Sixth Conference of the International

Association for Semiotic Studies

IASS-AIS

July 13-18, 1997

Guadalajara, Mexico

THE SEMIOTICS OF SCIENCE

By

Bradley S. Tice

Pacific Language Institute

Cupertino, California U.S.A.

Abstract

The lecture will examine the semantic and semiotic

values of current scientific terms and will be analyzed

in relation to the descriptive qualities of these terms as to what is actually transpiring. A selection of terms and concepts from the fields of chemistry, physics and biology will be examined in detail. The nature of the concept of the chemical bond will be explored in the field of chemistry and the notion of a wave will be examined in the field of physics. In biology the concept of an infinite quantity of a physical property or process will be examined in relation to natural constraints of the functions and size of those biological systems.

Introduction

It is necessary for terminology to have a descriptive quality that parallels an action of that description. The word 'description' is characterized by description or serving to describe. More to the point, it is concerned with a classification: a descriptive science (Dictionary, 1993). What words tell us about an action should accurately reflect the nature of that action, if it is to be of a descriptive nature. If the semantic value of a word; i.e. meaning, and the semiotics of that word; i.e. the symbol or sign used as a signifier of that word, are to give an accurate balance of the description of that action taking place, then the level of accuracy of the descriptive quality of that action must be very high or suffer in the translation.

In Charles Morris's <u>Foundations of the Theory of Signs</u>, he notes that civilization is dependent on signs and systems of signs, and that 'the human mind is inseparable from the functioning of signs – if indeed mentality is not to be identified with such functioning.' (Innis, 1985: preface). Morris continues with 'Indeed, it does not seem fantastic to believe that the concept of a sign may prove as fundamental to the sciences of man as the concept of the atom has been for the physical sciences or the concept of the cell for the biological sciences' (Innis, 1985: preface).

Vygotsky, the Russian semiotically influenced psychologist, thought that 'words' were paradigmatic in that a general reflection upon reality is the basic characteristic of words (Innis, 1985: preface). Words play a central part in the historical growth of consciousness as a word is a microcosm of human consciousness.

The linguistic sign, as pointed out by Saussure, is a double entity, in that a word is a fundamental unit in that it functions as a linguistic sign that units not a 'thing' and a 'name', but a 'concept' and a 'sound image' (Innis, 1985: 24). What Saussure means by 'sign' is the indissoluble union of the two components, an association, and is not a nomenclature (Holdcroft, 1991: 48). It is not 'a naming process only or a word list' but as Rene Thom would state: 'a foundation of images from a model appears like a manifestation of the irreversible character of a universal dynamics: the model ramifies into an image isomorphic with itself' (Innis, 1985).

Morris felt that the theory of signs has a two-fold relation to all other sciences in that it is both a science among the sciences and as an individual science, semiotics studies the things or properties of things in their function of serving as signs. But as every science makes use of and expresses its results in terms of signs, metascience, the science of science, must use semiotics as an 'organon' (Noth, 1995: 49).

Chemistry

"The theory of the chemical bond, as presented in this book, is still far from perfect. Most of the principles that have been developed are crude, and only rarely can they be used in making an accurate quantitative prediction."

From 1959 Preface from 1960 edition of "The Nature of the Chemical Bond" by Linus Pauling.

Pauling would develop the idea of the 'chemical bond' early on in his book "The Nature of the Chemical Bond" that is seminal in nature and defines many aspects of how we evaluate chemistry today.

"I knew a great amount of descriptive chemistry, and I could see how the shared pair of electrons could explain what the forces are that hold the atoms together. I could see that the first steps were being taken toward a real, systematic science of structural chemistry" (Judson, 1979: 73).

In defining 'interatomic distances and bond angles', Henry A. Levy uses crystallography as an example of the problem of chemistry as a whole. "When we look into this question, we come quite soon to the realization that at the points whose coordinates we've compiled, there are no physical objects. The points are in a strict sense totally imaginary. What they are averages, average positions. More than that, they are average positions of the center of gravity of diffuse clouds of scattering material, more diffuse for x-rays, less diffuse for neutrons. And what's more, these diffuse distributions are in constant motion. We are

pretty far from specifying exactly where something is"
(Milligan, 1980: 180).

Lewis notes the early work by Sir J.J. Thomas on the
definition of 'atomic interactions' by citing his article
"Forces between Atoms and Chemical Affinities"
(1914) in his seminal work on defining the 'chemical
bond' in his article "The Atom and the Molecule"
(Lewis, 1916) In this article, Lewis describes
differences in properties between polar and non-polar
types of atomic structures as being either an
'association' or that there is no 'association' between
them (Lewis, 1916: 763).

Lewis is specific when he defines this process as being
an 'assumption of the interpenetrability of the atomic

shells' and that 'an electron may form different atoms and cannot be said to belong to either one exclusively' (Lewis, 1916: 772). Lewis, who is always cited as being the first chemist to use or coin the notion and phrase 'a chemical bond' does, in fact, not see a 'bond' at all, but an 'association' between atomic and molecular structures.

Lewis would be the first to use a 'picture language' for the communicating of 'reactivity' information in chemistry by the use of a colon symbolic to represent the two electrons which act as the connecting link between the two atoms. "Thus we write Cl2 as Cl:Cl (Hont, Pietro and Hehre, 1984: preface, and Lewis, 1916: 777). Even his 'picture language', i.e. colon symbol (:) to represent a 'bond' re-enforces this misconception of a 'bonding' or 'attachment' between atoms or molecules as he has gone to great lengths to describe this process as a 'chemical association'.

It becomes apparent that the term 'chemical bond' should be replaced with another more accurate term for the process of atomic and molecular association, perhaps 'chemical association' or 'chemical fields' or even 'chemical spaces' but the term bond, that denotes a bonding or fixing of two or more elements as opposed to words that imply an association, pairing, or spacing of atomic or molecular structures, have a more accurate description of that process that the term 'chemical bond'.

Physics

Within conventional quantum theory are works only in (r,t)-space, i.e. wave and particle are considered as phenomena in (r,t)-space. Due to this conception, contradictions appear in the conventional quantum theory which can, however, be eliminated by means of the concept of complementarity. "According to the principle of complementarity, it is meaningless to talk about the physical properties (e.g. wave or particle) of quantum objects without precisely specifying the experimental arrangement which determines them" (Schommers, 1995: 147).

In other words, a phenomena (e.g. a wave or a particle) is always an observed phenomena (Copenhagen interpretation); without observation it is meaningless to talk about a phenomena. It is incorrect to regard a certain property of quantum objects as a property of the quantum object itself; rather, it is an attribute which must be assigned to both the quantum object and the experimental arrangement (Schommers, 1995: 46).

In other words, a wave is a series of 'point particles' and in a complementary state, a wave function is just another way of describing a similar process by the use of a particle.

Biology

The use of infinities in the applied sciences, i.e. biology, where abstract notions are incompatible with 'real' functional properties of those systems, the use of an infinity as a quantitative value is both inaccurate and an absurdity. We know that the universe is finite in size and that infinites only exist as conceptual and psychological impossibilities (Einstein, 1961: 112 and Benecerraf, 1983: 191).

When we use as a possibility of a unit of measure for a biological system as having infinite possibilities, we are ignoring a known constraint to such notions, in that all known biological systems, ecologies, and sub units, organism, have specific functional limits, in both the very small and the very large, and that what we are actual wanting to use is an indefinite, but bound, i.e. finite, quantity as a set function instead of an infinity, of which it could never be by the very nature of the physics of biology (Tice, 1997).

D'arcy Wentworth Thompson in his book <u>On Growth</u> <u>and Form</u> makes the point that all biological systems have an upper and lower limit to functioning and that these pose a 'natural constraint' to that organisms potential. In other words, each biological organism has a definite size requirement and that these requirements have definite functioning limitations (Newman, 1956: 996).

By incorporating an indefinite but bound quantitative value to these 'potentials' we have satisfied the need for an indefinite property, i.e. a variable, and that there is a limit, although not an exact placement for such a limit, a limit nonetheless, that reflects the constraints natural to these biological systems.

Summary

From these three examples, from the fields of chemistry, physics and biology, the real need to use scientific terminology that parallels those processes or properties of those actions thought to be carried out, are not only important as a descriptive nomenclature, but more importantly as a conceptual model of those processes as being self defining to that process.

References

American Heritage Dictionary 3rd Edition (1993) Boston: Houghton Mifflin Company.

Benecerraf, P. and H. Putnam (1983) Philosophy of Mathematics. Cambridge: Cambridge University Press.

Einstein, A. (1961) Relativity: The Special and the General Theory. New York: Wings Books.

Holdcroft, D. (1991) Saussure: Signs, Systems and Arbitrariness.Cambridge: Cambridge University Press.

Hout, R.F., Pietro, W.J., and Hehre, W.J. (1984) A Pictorial Approach to Molecular Structure and Reactivity. New York: John & Sons.

Innis, R.E. (1985) Semiotics: An Introductory Anthology. Bloomington: Indiana University Press.

Judson, H.F. (1979) The Eight Day of Creation. New York: Simon and Schuster.

Lewis, G.N. (1916) "The atom and the molecule" in <u>The Journal of the American Chemical Society</u>. January-June 1916.

Milligan, W.O. (1980) Proceedings of the Robert A. Welch Foundation Conference in Chemical Research. XXIII Modern Structural Methods. Houston, Texas.

Newman, J.R. (1956) The World of Mathematics. New York: Simon and Schuster.

Noth, W. (1995) Handbook of Semiotics. Bloomington: Indiana University Press.

Pauling. L. (1960) The Nature of the Chemical Bond. Ithaca: Cornell University Press.

Schommers, W. (1995) Symbols, Pictures and Quantum Reality. London: World Scientific.

Tice, B. "The limits of infinity" Presentation given at the Pacific Division American Association for the Advancement of Science (AAAS) 78th Annual Meeting. Oregon State University. June 22-26, 1997.

Appendix C

BRADLEY S. TICE

The Chemical Bond: A Misnomer?

By Bradley S. Tice

Pacific Language Institute

P.O. Box 2214

Cupertino, CA 95015-2214 U.S.A.

Telephone#(408)253-4449

Abstract

The term bond, used to denote a specific action by atomic and molecular chemical associations, is a poor semantic choice for what is thought to be occurring in chemistry and should be replaced by a more accurate term. Such terms as associative fields, mutual spaces and associative space are clearly a more accurate description of chemical association than is the word currently in use-bonding.

237

The Bond

The use of language to describe an action must, at least, come as close as possible to accurately describing what that action is doing. Precision in the use of a language parallels the accuracy of the action being described by that language. Terms maybe coined that reflect a general understanding of an action, that in time, are used without real reflection upon an accurate semantic model of that action. The term in question is the chemical word for atomic and molecular association: the chemical bond. The best example of this description of a bond, as it would relate to chemical bonding, is as follows; to bond, bound, a uniting or cementing force or influence by which a union of any kind is maintained (Simpson and Weiner, 1989: 380-381).

This would seem, at first glance, to correlate with the process associated with specific atomic and molecular association. Pauling defines a chemical bond as existing between two atoms or groups of atoms and that forces acting between them would lead to the formation of an aggregate with sufficient stability to be recognized by a chemist as an independent molecular species (Pauling, 1967: 5).

Bond Types

The types of bonds are covalent, ionic, metallic, and hydrogen. The covalent bond is a shared electron pair bond and was discovered by G.N. Lewis in 1916

(Pauling, 1970: 148). Ionic bonding is the result of Coulomb attraction of excess electric charges of oppositely charged ions (Pauling, 1960: 6). Metallic bonding is when the extra orbital is not occupied by an electron or electron pair in the neutral atom and is termed the metallic orbital and is the characteristic structure feature of metals (Pauling, 1970: 585). Hydrogen bonding is present when there is evidence of a bond and that this bond involves a hydrogen atom bonded to another atom (Pimentel and McClellan, 1960: 195).

A Review

If we were to reexamine the process associated with the chemical bond, the words used to accurately describe this process would fall into the category of association, pairings, mutual fields or shared spaces rather than joining, linking, bonding, or connecting. Because the language used in modern chemistry is the language of quantum mechanics, the importance of how the language is used becomes self-evident.

Importance of Language

Werner Heisenberg bases how we describe a process, and the language we use to describe it, as of primary importance (Heisenberg, 1930: 1-2). A more accurate term for the specific atomic and molecular associations would be one of the following: associative fields, mutual fields, or associative space. Each is a more accurate description of atomic and molecular association than is the current term in use-bonding.

References

Heisenberg, W. 1930 The Physical Principles of the Quantum Theory. Chicago: Dover Publications, Inc.

Pauling, L. 1967 The Chemical Bond. Ithaca: Cornell University Press.

Pauling, L. 1970 General Chemistry. San Francisco: W. H. Freeman and Company.

Pauling, L. 1960 The Nature of The Chemical Bond. Ithaca: Cornell University Press.

Pimentel, G.C. and A.L. McClellan. 1960 The Hydrogen Bond. San Francisco: W. H. Freeman and Company.

Simpson, J.A. and E.S.C. Weiner. 1989 The Oxford English Dictionary. Oxford: Clarendon Press.

Appendix D

BRADLEY S. TICE

The Veil of Valence

By Bradley S. Tice

The validity of the semantic worth of the chemical term valence as an accurate description of that chemical process is examined in this paper.

The concept of valence in the field of chemistry can be described as a process of force, an indentation in the charge density of an atom, an electronic orbit or that none of these explanations will be sufficient to explain the process. That no satisfactory explanation can account for the process of valency calls into question the language used to describe this Chimaera process. Theories of valence are not so much details of molecular structure as they are the basic principles of molecular formation.

The two main types of chemical bonds are ionic and covalent. The ionic bond is the result of the Coulomb attraction on the excess electric charges of oppositely charged ions. Pauling also mentions metallic bonds and the low-energy hydrogen bond as examples of other chemical bond types. In essence, the bonds in any molecule are simply the description of the electron distribution in it. Valence is the overlapping of atomic orbitals in a chemical bond.

Heisenberg states that language is incapable of describing the process occurring within atoms and that it would be difficult to modify our language so that it could describe these atomic processes because we can only describe actions that we have a mental picture of and this is not within the realm of possibility.

Although the processes occurring at the atomic and molecular level are difficult to properly enunciate in everyday language, the terminology should, at least, accurately reflect the actions postulated in the description of those processes. The word valence does not accurately reflect the processes occurring in the chemical bond.

References

1. Heisenberg, W. Across the Frontiers, New York, Harper & Row, Publishers, 1974., p.79.

2. Coulson, C.A. Valence, London, Oxford University Press, 1961, p.1.

3. Worrall, J. and I.J. Worrall, Introduction to Valence Theory, New York, American Elsevier Publishing Company, Inc., 1969, p.15.

4. Pauling, L. The Chemical Bond, Ithaca, Cornell University Press, 1967, p.5.

5. Pauling, L. The Nature of The Chemical Bond, Ithaca, Cornell University Press, 1960, p.5, 449.

6. Coulson, C.A. Valence, London, Oxford University Press, 1961, p.3.

7. Heisenberg, W. The Physical Principles of The Quantum Theory, Chicago, Dover Publications, Inc., 1930, p.11.

Appendix E

BRADLEY S. TICE

A History of the Chemical Bond

By Bradley S. Tice

Advanced Human Design

Cupertino, California U.S.A.

Abstract

A modern history of the chemical bond will be
explored starting with Lewis in 1916 and developing
the notion of a chemical bond by Pauling and ending
with current theories of the chemical bond.   A
chronological time line will present important
developments and figures in that development in the
history of the chemical bond.

Lewis Bond

The term 'bond' was coined in 1866 by Frankland for the combining power, or valency, of an atom. Although he was clear that such bonds "do not convey the idea of any material connection between elements of a compound" it was not until the advent of quantum mechanics that such 'affinities' become understood (Brock, 1992: 466).

Lewis was first inspired by Werner's <u>Neuere Anschauungen</u> of the early 1900's and the concept of 'co-ordination of atoms around a different central atom to form a positive or negative ion or neutral molecule' as a way of conceptualizing inorganic complexes (Brock, 1992: 468). Lewis would 'doodle' on the back of an envelope in 1902 'cubic atoms' in which the cubes shared, rather than exchanged, outer electrons along a common edge (Brock, 1992: 468-469). He would shelve these ideas for another decade until Bray and Branch published an article in 1913 replacing a polar and non-polar bonds with the total 'number' of polar or non-polar bonds; a total valence.

Lewis would comment on Bray and Branch's paper by stating "Since all electrons are alike, and presumably leaves no trail behind them, we cannot say that atom A loses an electron to atom B and atom C to atom D, but only that atoms A and C have each lost an electron and atoms B and D have each gained one" (Brock, 1992: 473). This would be the main idea behind the new quantum mechanics. The next year Thompson would develop a system were there was two tubes, two electrons, one from each atom.

At about the same time Alfred Parson, a graduate student of Lewis', suggested that the bonding be 'magnetic', not electrical, and published in a monograph in 1915. Lewis, using both Thompson's and Parson's ideas, developed the notion that where electrons were equally shared, the molecular exhibited no polar properties, but if one atom took an unfair share, then, the charge being unequally distributed, polarity would be produced.

Published as 'The atom and the molecule' in 1916, its lasting feature was the use of the colon [:] to represent paired electrons. Lewis' reputation was furthered still by both the publishing, in 1923, 'Valence and the Structure of the Atom' and the popularizing of his ideas by Irving Langmuir, especially in renaming the non-polar bond as a 'covalent bond' (Lewis, 1916: 777).

Valence Bond

By 1928 quantum physics had shown that the Schrodinger equation for one outer electron produced a symmetrical spherical orbital. When matched with the Heilter-London method, showing chemical bonding as overlapping orbitals, the chemical bond, was in essence, an electrostatic attraction between oppositely charged particles.

Linus Pauling would develop the valence bond theory into his seminal work, <u>The Nature of the Chemical Bond</u>, first published in 1939. <u>The Nature of the Chemical Bond</u> was the result of papers published over a period of ten years and as Pauling explains "I felt that I had an essentially complete understanding of the nature of the chemical bond by 1935"(Judson, 1979: 77).

Formulated in 1928, what would be called the 'Pauling rules' were the various relations among atoms that dictated the stable forms of crystalline substance that could be stated in a set of six principles (Judson, 1979: 76). Although not all original to Pauling, he did become valence bond theory's greatest spokesman.

Molecular Orbital

Developed by Hund and Mulliken and published in 1927-1928, the Molecular Orbital theory explained satisfactorily the more complex spectra of molecules by identifying, in the case of diatomic molecules like hydrogen, that vibrations due to the rotation of nuclei about one another, the oscillation in lengths of bonds between nuclei and the rotations and vibrations of the attached electrons.

This system codified spectroscopy, promoted the concept of correlating the orbitals of diatomic molecules with those of paired atoms and separate atoms. The one draw back to the Molecular Orbital theory was that Pauling was a better spokesman for Valence Bond theory than Mulliken for the Molecular

Orbital theory and that "Pauling made a special point of making everything sound as simple as possible" and was popular with the other chemists (Brock, 1992: 502).

Ultimately, the Valence Bond theory would decline because of the 'misconception' of resonance as operating in two different states at the same time. This was not helped by the fact that Pauling mentions in his book, The Nature of the Chemical Bond, that mathematically the bonds of AB take the form of A:-B, A-:-B, and A-:B (Brock, 1992: 504).

Summary

In summing the development of the concepts behind the various natures of the chemical bond, it is clear that ideas are not so much new as they are products of previous thoughts that have, over time, been refined and validated by empirical proofs.

Chemistry is as much a 'mind' game as chess, and has the added feature of being novel, if all the right cards are held, to produce a beautiful idea or concept. In looking back over the modern history of the chemical bond, it is clear that we are looking at a wonderful matrix of ideas and personalities as much as science. Let us hope that the future proves as interesting as the past.

References

Brock, W.H. (1992) The Norton History of Chemistry. New York: W.W. Norton & Company.

Judson, H.F. (1979) The Eighth Day of Creation. New York: Simon and Schuster.

Lewis, G.N. "The atom and the molecule" in The Journal of the American Chemical Society. Volume XXXVIII January-June 1916.

About the Author

Bradley S. Tice is currently Director and Institute Professor of Language and Linguistics at the Pacific Language Institute located in Cupertino, California

U.S.A. Mr. Tice has traveled around the world giving lectures, demonstrations, and papers on many aspects of language learning, language philosophy and the language of science.